ROUTLEDGE LIBRARY EDITIONS: PHILOSOPHY OF TIME

Volume 5

TIME DEVOURED

TIME DEVOURED

A Materialistic Discussion of Duration

EDMUND PARSONS

Routledge
Taylor & Francis Group
LONDON AND NEW YORK

First published in 1964 by Allen & Unwin

This edition first published in 2019
by Routledge
2 Park Square, Milton Park, Abingdon, Oxon OX14 4RN

and by Routledge
52 Vanderbilt Avenue, New York, NY 10017

Routledge is an imprint of the Taylor & Francis Group, an informa business

British Library Cataloguing in Publication Data
A catalogue record for this book is available from the British Library

ISBN: 978-1-138-39397-4 (Set)
ISBN: 978-0-429-40127-5 (Set) (ebk)
ISBN: 978-1-138-39402-5 (Volume 5) (hbk)
ISBN: 978-0-429-40122-0 (Volume 5) (ebk)

Publisher's Note
The publisher has gone to great lengths to ensure the quality of this reprint but points out that some imperfections in the original copies may be apparent.

Disclaimer
The publisher has made every effort to trace copyright holders and would welcome correspondence from those they have been unable to trace.

TIME DEVOURED

A Materialistic Discussion of Duration

EDMUND PARSONS

London
GEORGE ALLEN & UNWIN LTD
RUSKIN HOUSE MUSEUM STREET

FIRST PUBLISHED IN 1964

PRINTED IN GREAT BRITAIN
in 10 pt. Juliana type
WILLMER BROTHERS AND HARAM LIMITED
BIRKENHEAD

DEDICATION

How, oft in after years when I am gone,
Others shall seek thee for thy beauty's grace,
And vacant in thy heart shall find the place
That I must leave when my brief stage is done;
How lovers new shall look upon thy face,
Viewing refulgent there their world's high sun,
Wooing thy ways until thou shalt be won
To cancel from thy thought my ancient trace;
This do I know. Yet is my vantage one
That all thy future never can efface;
For I have loved thee in the fairest space
Of thy fair youth, so that there shall be none
Who seeth thee as I shall see thee ever:
Lovelier than thou art now, shalt thou be never.

ACKNOWLEDGMENTS

My grateful acknowledgments are due to the following authors and publishers: to Messrs Edward Arnold for permission to quote an extract from *Physical Similarity and Dimensional Analysis* by Dr W. J. Duncan; to Messrs Basil Blackwell for permission to quote several passages from Ludwig Wittgenstein's *Philosophical Investigations;* to Messrs J. & A. Churchill for permission to quote extracts from the *Ciba Foundation Symposium on Extrasensory Perception;* to Messrs William Heinemann for permission to make a quotation from *The Black Cloud* by Fred Hoyle; to Messrs Longmans, Green & Co., London, and The Ronald Press Co., New York, for permission to quote a passage from *Christian Theology and Natural Science* by Dr E. L. Mascall; to Messrs Methuen & Co. for permission to quote two extracts from Einstein's *The Meaning of Relativity;* to Messrs Routledge & Kegan Paul for permission to quote extracts from *The Mind and Its Place in Nature* by Prof. C. D. Broad, from *Analysis of Perception* by Dr J. R. Smythies, and from Wittgenstein's *Tractatus Logico-Philosophicus,* the translation used being the old one originally published in the International Library of Psychology, Philosophy and Scientific Method in 1922—a new translation by D. F. Pears and B. F. McGuinness was published by Messrs Routledge & Kegan Paul in 1961; to Messrs Martin Secker & Warburg for permission to quote four extracts from George Orwell's *Nineteen Eighty-Four;* to Messrs Sheed & Ward for permission to quote several extracts from F. J. Sheed's translation of *The Confessions of St Augustine;* to The Syndics of the University Press, Cambridge, for permission to quote passages from *The Evolution of Physics* by Albert Einstein and Leopold Infeld, from *Scientific Inference* by Sir Harold Jeffreys, and from Erwin Schrödinger's *Space-Time Structure;* and, finally, to The Society of Authors as the Literary Representative of the Estate of the late James Joyce for permission to make three quotations from *Finnegans Wake,* and to J. M. Whittaker, Esq., F.R.S., for kindly allowing me to quote two passages from Sir Edmund Whittaker's *From Euclid to Eddington.*

Here, too, I express my sincere thanks to Lt.-Col. R. E. H. Waring for his advice on ballistics, to the librarians of the British Council in Rome for their courtesy, and to my wife Giulietta for her patience.

EDMUND PARSONS

L'ALLEGRO

This is a gift that I have, simple, simple; a foolish extravagant spirit, full of forms, figures, shapes, objects, ideas, apprehensions, motions, revolutions. These are begot in the ventricle of memory, nourished in the womb of pia mater, and delivered upon the mellowing of occasion. But the gift is good in those in whom it is acute, and I am thankful for it.

Love's Labour's Lost, IV,2

IL PENSEROSO

Abstruse thought and profound researches I prohibit, and will severely punish, by the pensive melancholy which they introduce, by the endless uncertainty in which they involve you, and by the cold reception which your pretended discoveries shall meet with when communicated.

HUME *Enquiry Concerning Human Understanding* I

CONTENTS

I

ELYSIUM

'Drive your cart and your plough
over the bones of the dead.'
William Blake, *The Marriage of Heaven and Hell.*

1

About time, much nonsense has been written, most of it in the twentieth century. The concept has always been obscure, but metaphysical mathematicians, block-universe enthusiasts and precognition theorists have bedevilled it to absurdity. The purpose of this book is to make it clear.

It was Plato, as usual, who started the discussion on the wrong foot. To him, undoubtedly, belongs 'the great merit of distinguishing Time as it is in itself from our awareness of Time and from events in Time'. The commendation is Dr W. R. Inge's,[1] whose library evidently contained the *Encyclopaedia of Religion and Ethics,* where another Platonist, Professor C. D. Broad, originally conferred it: 'Plato . . . has the merit of distinguishing time from what is in time.'[2] Thus illuminated, Dr Inge was comforted to find that 'there is no difficulty in thinking of empty Time or of empty Space, any more than in thinking of a blank sheet of paper'.

By coincidence of contraries, the Marxist scriptures commit orthodox dialectical materialists to belief in the same Platonist phantom: 'And time in which no recognizable changes occur is very far removed from *not* being time,' Engels contended; 'it is rather *pure* time, unaffected by any foreign admixtures, that is, real time, time *as such.* In fact, if we want to grasp the idea of time in all its purity, divorced from all alien and extraneous admixtures, we are compelled to put aside, as not being relevant here, all the various events which occur simultaneously or one after another in time, and in this way to form the idea of a time in which nothing happens. In doing this, therefore, we have not let the concept of time be submerged in the general idea of being, but have thereby for the first time arrived at the pure con-

[1] Mysticism in Religion, V.
[2] E.R.E., art. 'Time'.

cept of time.'[3] It is typical of that intellectual subserviency which is the bane of dialectical materialism that the Marxist philosopher Maurice Cornforth, after stating correctly in the 1946 edition of his *Science versus Idealism* that 'time (and space) are not like a box, in which events are placed but which could just as well exist empty,'[4] and acknowledging 'with gratitude'[5] Professor J. B. S. Haldane's 'very valuable suggestions' on the subject, omitted his whole discussion of time from the enlarged 1955 edition which, suitably replenished with quotations from Marx, Engels, Lenin, Stalin and Mao Tse-tung, was 'especially indebted to the criticisms of my previous work contained in reviews appearing in the British journal *Communist Review*, the American *Political Affairs*, and the Soviet *Bolshevik* and *Problems of Philosophy*.'[6]

Like other Platonist balloons, time the moving image of eternity was first punctured by Aristotle. 'For time is just this,' he said, 'the counting of motion in respect of before and after. . . . Time does not exist without motion. . . . It is an aspect of motion.'[7] And: 'There is no such thing as motion over and above things, for there is nothing over and above them. . . . To be in time does not mean to coexist with time any more than to be in motion means to coexist with motion.'[8] Broad's complaint that 'there seems no reason to think that Aristotle was really clear as to the distinction between time and motion,'[9] which echoes the argument of Plotinus, who derived it from Plato, that 'motion time cannot be, since motion takes place in time,' merely indicates the distance that separates much modern philosophy from Aristotelian common sense.

St Augustine followed Plato, and consequently found time an *implicatissimum aenigma*: 'From infancy I came to boyhood, or rather it came to me, taking the place of infancy. Yet infancy did not go, for where was it to go to? Simply it was no longer there. . . . O Lord my God, I measure time and I do not know what it is that I am measuring!'[10] His puzzlement has echoed down the centuries: 'What, then, is time? If no one asks of me, I know: if I wish to explain to him who asks, I know not.'[11] But the main stream of Scholastic thought after St Thomas Aquinas followed Aristotle, and the old definition prevailed until Descartes operated that dichotomy between the thinking subject and the external universe which, whilst it left science free to advance unencumbered by the doctrine of teleologically active

[3] *Anti-Dühring*, V.
[4] 1946 edition, XII, 2.
[5] Ibid., 'Acknowledgments'.
[6] 1955 edition, 'Foreword'.
[7] *Physics*, IV, 11, 14.
[8] Ibid., III, 1; IV, 12.
[9] Loc. cit.
[10] *Confessions*, I, viii; XI, xxvi.
[11] Ibid., XI, xiv.

forms, left philosophy struggling to reunite its sundered world, in which effort it is still largely engaged.

Time, which hitherto had been an objective measure of motion, became for the new metaphysics a subjective experience of duration.[12] 'Descartes,' Hegel proclaimed, 'is a hero; he made a fresh start and rediscovered the true ground of philosophy, to which it returned after an aberration of a thousand years.'[13] Now Hegel's thousand years take us back beyond St Thomas to the period when the Schools were dominated by the teaching of St Augustine, from whom Descartes in fact obtained his 'Cogito, ergo sum'. It is Augustine and not Aquinas who is rightly regarded as the father of modern philosophy; and behind Augustine stands Plato.

Smaller doses of Platonism during the Scholastic period had been fairly well tolerated; but the drastic injection of it in the degraded form of the inert substances *res cogitans* and *res extensa* split the Scholastic universe alive with potentiality into a closed inner world of the mind impotent to contemplate other than its own ideas and a cold outer world of lumps of dead matter moving in a vacuum. The only bridge between the two worlds was the will of God, and it was soon to be withdrawn. On one side of the gulf philosophy was left to inflate its concepts into unreality; on the other, physics formulated the mechanics of the material clockwork. For philosophy, the world was to become insubstantial; for science it was already inertial.

The new metaphysics was transmitted to science by Descartes' friend Pierre Gassendi, memorable for his ambition to become another Aquinas by changing the philosophical basis of Christianity, the system he proposed for the purpose being Epicureanism. Whilst for the progressive derangement of philosophy he sponsored the revival of the word 'ideas', which Aristotle had abandoned to the Platonists, to denote the 'immediate objects of knowledge' (that is, not external objects but more or less what are nowadays called 'sense-data'), for the future confusion of physics he resurrected the Epicurean doctrine of the void that, in opposition to Descartes' own view of space as a mere attribute of matter, was adopted by Newton.

'Absolute Space,' Newton decreed, 'in its own nature, without relation to anything external, remains always similar and immovable.' His definition of time was parallel: 'Absolute, True and Mathematical Time,' he wrote, 'of itself, and from its own nature, flows equably without relation to anything external, and by another name is called

[12] For the *Encyclopaedia Britannica*, 14th edition, it still is: 'TIME, the general term for the experience of duration.'

[13] *History of Philosophy*; quotation translated from the appendix of a Larousse edition of Descartes' *Discours de la Méthode*.

duration: relative, apparent and common time, is some sensible and external (whether accurate or unequable) measure of duration by the means of motion, which is commonly used instead of true time; such as an hour, a day, a month, a year. . . . It may be, that there is no such thing as an equable motion, whereby time may be accurately measured. All motions may be accelerated or retarded, but the true or equable flowing of absolute time is liable to no change. . . . And so, instead of absolute places and motions we use relative ones; and that without any inconvenience in common affairs; but in philosophical disquisitions we ought to abstract from our senses, and consider things themselves, distinct from what are only sensible measures of them.'[14]

This is Platonism with a vengeance. But it would be a mistake to suppose that Newton adopted it on philosophical grounds; he did so, rather, on physical ones, his new physics itself demanding a new metaphysical foundation. Cartesian metaphysics was as vital for the generation of science as it was fatal for the degeneration of philosophy.

The ideal purpose of science, although scientists are strangely bashful about this nowadays, is the construction of a conceptual system that shall constitute *knowledge* of the world. Its practical purpose is control of the world, material and intellectual power. But before science can begin to know and control, it must classify and measure, and the science of mechanics is based upon measurements of simultaneity and position. Without a rigid spatial framework, preferably absolute, within which position could be located, and an 'equably flowing' background of absolute time against which simultaneities could be set, physics could hardly start.

Such a Platonist metaphysics had, at that stage of science, no disadvantage. The basic doctrine of Platonism, that what changes cannot be real, leaves science unperturbed; its dealings are not with noumena but with phenomena, the pejorative significance of the very word 'phenomena' being ignored. For the Platonist, phenomena are subjective shadows; for the scientist, they are objective manifestations of nature. What a Platonist metaphysics gives to science is an absolute framework for the concepts whereby phenomena can be reduced to intelligibility and thence to subjugation. To Newton it gave absolute space and absolute time; to Einstein it gave the absolute space-time continuum.

Equipped with the Newtonian absolutes, leaving the presumably immaterial microcosm of the 'mind' to be reckoned with later, science proceeded to the conquest of power over the material world. Philoso-

[14] *Principia*, Scholium to Definition VIII.

phy, impotent to establish the very existence of that world, was first to doubt and finally to deny that it existed at all.

Leibniz, perhaps because he dissolved the dualistic universe of Descartes into a plurality of independent microcosms, was still able to declare time to be merely relative, 'the order of successive existents' in the same way as space was 'the order of co-existents', and his rear-guard stand against the advancing doctrine of Newton is one of the memorable battles of philosophical history. Also Berkeley and Hume directed their shafts against it, but for them the succession of existents had become a succession of ideas. This led to Kant, in whom the futility of metaphysics stood self-confessed: a world external to mind was not denied as in Berkeley, nor was it contemptuously relegated to the realm of faith as in Hume; the conception of its existence was a rational necessity, but it was totally inaccessible to any possibility of experience.

Even an unknowable external world was felt by such subsequent thinkers as Fichte to be a deplorable compromise with materialism. Kant himself assumed it because without some influx from reality the subjective world would have consisted of empty concepts, a 'bloodless dance of the categories'.[15] However, the world of experience was not produced by that external reality but by the purely mental forms of perception and conception applied to the very raw material coming in. The subjective matrices of perception were space and time; that is to say, time was a necessary aspect of our experience, but it had no existence in external reality. The unknowable world of 'things-in-themselves' was spaceless, timeless and insubstantial; the empirical world was created out of chaos by the thinking subject.

When Kant had finished, only two possibilities remained for intellectual activity: science could scrutinize the gigantic reflection of its own self in the empirical world, and philosophy could analyse the rational processes which produced that reflection. Kant believed that this offered science its proper metaphysical foundation; but the Greek gift was generally declined until Sir Arthur Eddington dragged it into the fortress a century and a half later.

With Hegel the philosophical nightmare reached its climax; the sleeper could awake, if he wanted to. The rational processes that produced the visible universe were not in the mind of man; they were in the mind of God Almighty. Brahma begot the cosmos upon himself in inexhaustible superfetation.[16] The whole world was a theophany, the necessary foaming forth of Infinite Being spatially

[15] as William James said somewhere.

[16] Or, as the Vedas have it, by incest with his daughter Vac (Speech)—the generation of the idealist's world!

and temporally in an objective externalization eternally comprehended as the self-consciousness of its own Absolute Spirit. There was no problem of knowledge, no Cartesian gulf between mind and world, subject and object. The mind and its objects were in immediate unity as inseparably interrelated phenomena of that absolute spiritual reality from and within which they had their differentiated existence. Similarly, the rational opposition of time and space arose from their being dialectically one: 'This immediate unity of space and time is already the ground, through which they are.' The anticipation of Minkowski's four-dimensional continuum was still more explicit in Schelling: 'We can define space as arrested time, time as flowing space.'[17]

Philosophy could soar no higher. Although Hegel's gigantic grasp of reality as process offered science a key not to the mere description of the world but to its comprehension, his system as such was scientifically useless. Leaving the further speculations of Pure Reason to Schopenhauer, ostensibly making do with a metaphysical motley strung together from Cartesian tatters but actually using materialism as its unacknowledged heuristic principle, science pressed on with its proper business of questioning sensible existence in order to obtain sensible answers.

It was, at that period, in trouble with the aether. This ghost of Stoic cosmogony had lingered on in the background for two thousand years. Even Newton, although his corpuscular light needed no medium, and 'to make way for the regular and lasting motions of the planets and comets it is necessary to empty the heavens of all matter', left the 'most subtle spirit' of the aether unexorcised.[18]. It remained the vehicle for the longitudinal light-waves of his neglected rival Huygens. But in 1804 Thomas Young revived the wave theory as the only possible explanation of his optical experiments, and in 1817 he discovered that light-waves were not longitudinal after all, but transverse, which set scientists the problem of conceiving a medium dense yet elastic enough to withstand the strain of their vibration. With the undisputed collapse of Newton's corpuscular theory after Foucault's refraction experiments in 1850 and the incorporation of light into the whole range of electro-magnetic radiation by Maxwell in 1873, the seeds of confusion sown by Gassendi two centuries earlier reached full growth. The aether

[17] Quotations in E.R.E., art. 'Space'.

[18] Despite the 'hypotheses non fingo' of his *Principia* (1713) he was wondering in 1717 whether it might not explain the force of gravity. (See Mr H. G. Alexander's edition of *The Leibniz-Clarke Correspondence*, pp. 171-173, extracts from Newton's *Opticks*.)

became thicker and thicker, and scientific ingenuity exhausted itself in endeavours to conceive its nature.

Then, in 1905, Einstein cast his bomb, and the aether exploded. In 1908 Minkowski blew his petrifying trumpet-blast: 'Henceforth space and time as separate realities fade away as shadows and only a kind of union of them both has absolute existence!' And the entire cosmic machinery ground to a dead stop.

2

Spellbound within the magic sphere of the space-time continuum the world was transformed and the weirdest things happened, in so far as anything at all could be said to happen in a world where becoming was abolished. Multiplied by the velocity of light and the square root of minus one, time became the fourth dimension. The curvature of space involved an inscrutable mystery: though finite, its centre was everywhere and its boundary nowhere. A moving body contracted in the direction of its motion, shrinking as its velocity increased until at the speed of light its length became zero and it vanished; yet at the same moment its mass became infinite. All clocks on such a body went slower and slower, until at 186,000 miles a second time stood still. Spacemen travelling in a photon-rocket to the Taurus cluster and back would return unaged to earth to find, like the Seven Sleepers of Ephesus, a brave new world; by extending their circuit to an outlying nebula, the immortal youth of their landfall would be the last desolation of H. G. Wells's Time Machine. Popular imagination was certainly gripped, even though it was difficult for it to grasp anything.

It is all very well for physicists, deploring popular misconceptions of relativity theory, now to tell us that no metaphysical significance should be attached to its pronouncements, that they are merely a matter of geometry 'and there is no metaphysics of geometry any more than there is a metaphysics of the multiplication table.'[19] These are the reconsiderations of maturity; the proclamations of revolutionary youth resounded in more startling tones.

For such general misunderstanding of the significance of relativity theory as still prevails, relativity theorists themselves are largely to blame. If the theory has been popularly misunderstood, it is because it was popularly misrepresented in the first place. Such elucidation as the physicists eventually supplied was certainly occasioned by a genuine desire to put matters in their proper perspective, but it was also due to criticism by those who were not physicists. We are

[19] Sir Edmund Whittaker, *From Euclid to Eddington*, I, 3.

not quarrelling with mathematical equations, but with English expressions, and it will not do at all to be told that the misunderstanding was a deficiency of intelligence when it was a deficiency of intelligibility. Whether their mathematical training led them to scorn language as a medium of rigorous thought so that they did not care to attempt precise expression in it, whether they really believed what they said or whether they were just straining themselves to strike us with wonder at their discoveries, the fact remains that when relativity theorists translated their differential equations into English, the result was grossly misleading.

Contrary to the general belief, there is nothing extraordinarily difficult to understand about the theory's essential significance. What does make the effort to understand it so intellectually exasperating is the paradoxical manner in which it has been expounded.

Typical of such misteaching is the book entitled *The Evolution of Physics* written by Einstein himself in collaboration with Leopold Infeld. It is practically impossible, for example, to read the chapter headed 'Time, Distance, Relativity' and realize that the contraction of the moving rods and the slowing down of the moving clocks are only apparent changes and not real ones. It is practically impossible (unless, of course, one knows it already) because it is never stated that these changes *are* only apparent. On the contrary, everything is done to emphasize that they are real: 'If the velocity of light is the same in all c.s.,[20] then moving rods must change their length, moving clocks must change their rhythm, and the laws governing these changes are rigorously determined. There is nothing mysterious or unreasonable in all this. In classical physics it was always assumed that clocks in motion and at rest have the same rhythm, that rods in motion and at rest have the same length. If the velocity of light is the same in all c.s., if the relativity theory is valid, then we must sacrifice this assumption. It is difficult to get rid of deep-rooted prejudices, but there is no other way.'[21]

Then, in the chapter 'The Time-Space Continuum', we are told that the representation of *motion* as *static* is 'a more convenient and *more objective* picture of reality'.[22] Again, 'from the point of view of the relativity theory the *static* picture is the more convenient and the *more objective*.'[23]

More convenient for the theory of relativity, granted. But more *objective*? We realize that what Einstein here means by 'objective' is something different from what the word means in the dictionary. What *does* he mean? He explains himself immediately:

[20] coordinate systems.

[21] Scientific Book Club edition, pp. 195-196.

[22] Ibid., p. 217; my italics.

[23] Ibid., p. 220.

'Even in the relativity theory we can still use the dynamic picture if we prefer it. But we must remember that this division into time and space has no objective meaning since time is no longer "absolute".'[24] It transpires that only that which is absolute can be objective. The Platonism is obvious.

Admittedly, the meaning of 'objective' is hardly fixed; it has boxed its compass once already,[25] and is now oscillating so unsteadily that it seems likely to do so again. Modern scientific writers avoid it. As Sir Harold Jeffreys observed, 'The final stage was achieved by Eddington, who regarded the laws of physics as subjective and psychology as objective. In the circumstances, the less the words are used the better.'[26] Even before Eddington, the dubious value of 'objective' enabled idealists to maintain that it was wrong to make it equivalent to 'external', and that it should properly be applied to whatever could be thought or about which statements could be made, thus conferring objectivity and hence reality upon such absurdities as square circles.[27] But this argument, however disreputable, is irrelevant to the use of 'objective' made by Einstein, because such idealist realities as square circles are certainly not objects in the physical world of which relativity theory professes to give an objective representation.

Nevertheless, as long as dictionaries are admitted to establish the present meaning of words, such statements as that it is objective to regard motion as static are inadmissible. '*Objective*: . . . Used of the existence or nature of a thing *as an object of consciousness* (as distinguished from an existence or nature termed *subjective*). . . . That is or belongs to what is presented to consciousness, as opposed to the consciousness itself; that is the object of perception or thought, as distinct from the perceiving or thinking subject; hence, that is, or has the character of being, a "thing" external to the mind; real.'[28]

In this conflict between Einstein and the *Oxford English Dictionary*, we can only echo Leibniz: 'And it will be a hard matter to put a justifiable sense upon this word, according to the use Sir Isaac Newton makes of it. Though the question be about the sense put upon that word by Sir Isaac Newton, and not by Goclenius; yet I am not to blame for quoting the philosophical dictionary of that

[24] Idem.
[25] The Scholastic 'objective' was roughly our 'subjective', and vice versa.
[26] *Scientific Inference*, 9, 8.
[27] E.R.E., art. 'Object, Objective'.
[28] *Oxford English Dictionary*; v. also Webster's *New International Dictionary*.

author, because the design of dictionaries is to show the use of words.'[29]

What Einstein wanted, of course, was to make 'objective' mean 'not relative'. But there is a word already in the dictionary for that: 'absolute'. Why, then, did he not use it? The answer takes our realization of his meaning a step further, and it is this: *Because he also wanted to make his space-time continuum out to be a physical object.*

It is characteristic of physical objects that they *do* have different appearances from different viewpoints. Relativity theory shows the same to be true, under certain conditions, of their spatial and temporal relations. This does not stop our observation of them being objective; it only makes it relative. But according to Einstein, if we could somehow see things from all round at once, that would make our observation of them 'more objective'; and the genius who produced a 'picture' of them from such an absolute view would consequently be justified in calling it a 'more objective picture'.

What Einstein was telling us, in fact, was not only that space and time (or, shall we rather say, spatial and temporal intervals) are relative to their observers, already a conclusion of great importance, but that *when we observe spatial and temporal intervals we are really, from our various subjective viewpoints, observing an absolute physical object, the four-dimensional space-time continuum.* In *The Meaning of Relativity* he explicitly defined its objective physical reality: 'Just as it was necessary from the Newtonian standpoint to make both the statements, *tempus est absolutum, spatium est absolutum,* so from the standpoint of the special theory of relativity we must say, *continuum spatii et temporis est absolutum.* In this latter statement, *absolutum* means not only "physically real", but also "independent in its physical properties, having a physical effect, but not itself influenced by physical conditions".'[30]

That something was wrong with this conception of the space-time continuum as an *independent* physical entity was immediately observed by Einstein himself: 'it is contrary to the mode of thinking in science to conceive of a thing[31] (the space-time continuum) which acts itself, but which cannot be acted upon;'[32] and he tried to remedy this in his subsequent General Theory by the assumption that the geometrical properties of the continuum depend upon the gravitational fields of the masses it contains. But it still remained

[29] Leibniz' Fourth Paper in *The Leibniz-Clarke Correspondence.*

[30] 4th edition, p. 54.

[31] ?—?—?—?

[32] Ibid.

mysterious. Dr G. J. Whitrow, for example, complained that it was never clear whether the presence of matter caused the curvature of the continuum, or whether the curvature of the continuum caused the existence of matter, and he found himself reminded of a remark by Bishop Barnes that the astonishing thing about Einstein's equations was that they appeared to have come out of nothing.[33]

The truth of the matter is, that the mystery of Einstein's space-time continuum was simply due to the contradiction inherent in his definition of it as an absolute physical entity. He never seems to have realized that it was not a physical entity at all, that it was a complete abstraction, and so could not possibly have either physical properties or physical effects. All his life he continued to think of it as a physical object, an absolute *thing*.[34]

No criticism from any quarter, least of all from this one, can detract an iota from Einstein's greatness as a genius of physics, and philosophers can, or should, be grateful for his stand along with Schrödinger and De Broglie against the indeterminism of orthodox quantum theory.[35] But before the true meaning of his work can be understood, it must be given its proper metaphysical background. Physics without metaphysics is practicable, but unintelligible. So far from metaphysical propositions being meaningless to physics, physics is meaningless without them.[36]

Unfortunately, starting from Einstein himself, the metaphysics that accumulated behind his physics and largely determined the manner of its presentation was deplorable. 'The constructive achievement of Einstein was to combine what was known of gravitation and light into a single scheme and to find the appropriate generalization of Laplace's equation when the finite velocity of light is taken into account. This was a great feat of scientific imagination, and I think it would be more appreciated if the bad philosophy associated with the theory was cleared away so that the actual achievement can be seen.'[37]. Although among philosophically competent physicists this judgment is now a commonplace, the belief that Einstein provided a scientific basis for the metaphysical fusion of time and space into a

[33] *The Structure of the Universe*, pp. 73-74.

[34] See H. G. Alexander's excellent summary of the theory in his introduction to *The Leibniz-Clarke Correspondence*.

[35] the 'Copenhagen' school of Bohr and Heisenberg.

[36] Cf. the paper 'On Philosophical Arguments in Physics' by Prof. S. Körner in *Observation and Interpretation*. Proceedings of the IXth Symposium of the Colston Research Society.

[37] Sir Harold Jeffreys, *Scientific Inference*, 8, 8.

four-dimensional absolute reality is still prevalent even among those who should know better.

Despite, therefore, the disclaimers of relativity theorists, their protests that mathematical equations are not to be translated literally, the fact remains that they were so translated, and badly, so that to prick the bubbles inspired by those equations is still a necessary task. Once these are reduced to their proper substance, the metaphysical air-castles constructed from them collapse into so much spindrift. Time the fourth dimension is the most unsubstantial of them all.

3

Before our eyes the world unfurls its myriad forms in quenchless change; we move upon the quick substance of the hurtling earth; here we are born, we breathe and grow and do things before we die. But this is maya, illusion: in the 'more objective' world of the space-time continuum where dynamics is reduced to statics, nothing is born, nothing breathes, nothing grows, nothing is achieved, and we are 'four-dimensional worms'.[38] The block universe stands a frozen absolute wherein past and future exist fast-bound, extended for eternity along the dimension of time undone.

First, it must be remarked that, like 'objective' as used by Einstein, the word 'dimension' as used by many physicists has a meaning which is not to be found in the *Oxford English Dictionary*. It is not, that is to say, 'Measurable or spatial extent of any kind, as length, breadth, thickness, area, volume', nor is it 'A mode of linear measurement, magnitude or extension, in a particular direction; usually as co-existing with similar measurements or extensions in other directions,' which is given as the geometrical definition. 'Here the notion of measurement or magnitude,' the Dictionary adds, 'is commonly lost, and the word denotes merely a particular mode of spatial extension. Modern mathematicians have speculated as to the possibility of more than three dimensions of space.'[39]

The physicists' use of the word does, however, correspond more or less to definitions given in Webster's *New International Dic-*

[38] Eddington, *The Nature of the Physical World*, pp. 42, 53, 92. Also Lord Russell, around 1924, was talking of 'spatio-temporal tubes'; v. 'Logical Atomism' in *Logic and Knowledge*, ed. Marsh. Cf. William Blake: 'I am your rational power, O Albion, and that human form You call divine is but a worm seventy inches long.' (*Jerusalem*, 33.)

[39] The 'D' volume was published in 1897. The Supplement (1933) has no relevant addition to make, nor is there any significant modification of the above definitions in the 1955 edition of the *Shorter Oxford English Dictionary*.

tionary:[40] 'The degree of manifoldness of an aggregate as fixed by the number of determinations or conditions necessary and sufficient to distinguish any one of its elements from all others. . . . In general, a magnitude is of varying dimensions according to the elements of which it is made up. . . . The manifoldness or degree with which the fundamental units of time, length and mass enter into the units of other physical quantities. Thus, since the unit of velocity varies directly as the unit of length and inversely as the unit of time, the dimensions of velocity are said to be length divided by time.' That is to say, velocity is two-dimensional. In the same way, viscosity is three-dimensional, being usually formulated as mass divided by length and time; density is four-dimensional, being mass divided by length cubed; energy is five-dimensional, being mass multiplied by length squared divided by time squared; and power is six-dimensional, mass multiplied by length squared divided by time cubed.[41]

The word 'dimension' as used by physicists, therefore, appears to be an Americanism, and what it signifies is a scientific concept expressible as a measurable quantity that can enter as an element in the formulation of more complex quantities, and *not* a physical extension, *not* a direction along which things exist, *not* a lineament of shape at all. It was disgust with this ambiguity that led an eminent writer on dimensional analysis to reject the word altogether: 'Throughout the book I have used the phrase *measure formula* as a synonym for *physical dimensions* and I consider that "measure formula" is much to be preferred because it is a simple and correct expression of the entity. There is a metaphysical smack about "physical dimensions" which has misled and bamboozled many.'[42]

The astonishing thing is that physicists themselves, who ought to have known what they were talking about if anybody did, were the first to be bamboozled. Besides Minkowski and Einstein, it seems that also De Broglie, for example, believed that the conception of the space-time continuum necessarily implied in physical reality the negation of all becoming.[43]

[40] 1926 edition.

[41] Modern dimensional analysis does not regard the units of time, length and mass as in any way intrinsically fundamental; the use of certain units as primary is a matter of convenience and convention. Cf. Dr G. W. Scott Blair, *Measurements of Mind and Matter*, pp. 43-50; also Dr W. J. Duncan, *Physical Similarity and Dimensional Analysis*, 2.3, 2.4.

[42] Duncan, op. cit., preface.

[43] See the quotation from his book *The Revolution in Physics* in Dr E. L. Mascall's *Christian Theology and Natural Science*, V. 1.

But it is not easy to think consistently in four dimensions, even though the four-dimensional picture is asserted to be 'more convenient', and relativity theorists can often be caught napping in this respect. The space-time world is by definition completely motionless, and this reduction of living creation to a rigid cadaver undoubtedly has scientific charm. Nevertheless, it remains so fundamentally unreal that scientists themselves sometimes talk about light-rays being propagated in it, or describe consciousness as crawling along its 'world-lines'.[44] Professor H. Weyl, for example, at home though he was with vector-potentials, was yet able when he contemplated the space-time continuum to commit a blunder like this: 'However deep the chasm may be that separates the intuitive nature of space from that of time in our experience, nothing of this qualitative difference enters into the objective world which physics endeavours to crystallize out of direct experience. It is a four-dimensional continuum, which is neither 'time' or 'space'. *Only the consciousness that passes on in one portion of this world experiences the detached piece which comes to meet it and passes behind it* as history, that is as a process that is going forward in time and takes place in space.'[45]

Obviously, to allow anything to 'pass on' in the space-time continuum or anything else to 'come to meet it' there, disrupts the precious immobility of the entire system by requiring another time in which these movements can be performed, and so flings wide the gates of inanity for Serialism and its kind to come stalking in.

'But there are worthies a-coming will speak their mind in some other sort.'[46] Here, an' it please you, is Dr J. M. J. Kooy, professor of theoretical physics at the Royal Military Academy, Breda, Holland —a chance encounter, but to our purpose. Prof. Kooy is quite clear about the space-time manifold allowing no movement of any kind, and has no hesitation in declaring cause and effect, speed and acceleration, change and destruction, the difference between past, present and future, between the living, the dead and the not-yet-born to be subjective illusions. Since, therefore, our later brain-states already exist, and having 'fully confirmed' the occurrence of precognition by 'my own introspective studies of my dreams', he easily provides a scientific explanation of it complete with a mathematical appendix.[47]

[44] Cf. Prof. J. J. C. Smart's stricture upon these fallacies in his paper 'The River of Time' (*Essays in Conceptual Analysis*, ed. Flew).

[45] My italics. Quoted without comment in Whitrow's *The Structure and Evolution of the Universe*, pp. 85-86; no reference given.

[46] *Love's Labour's Lost*, V. 2.

[47] 'Space, Time and Conciousness', *The Journal of Parapsychology*, December 1957.

As Leibniz truly said, 'These are imaginations of philosophers who have incomplete notions, who make space an absolute reality. Mere mathematicians, who are only taken up with the conceits of imagination, are apt to forge such notions.'[48]

Pleasantries apart, it must now be asked why, since in scientific terminology a dimension is not a physical extension, not a direction along which things exist, not a lineament of physical shape, but *any* kind of measurable quantity that it is scientifically convenient to regard as an elementary constituent of other kinds of quantity, time is represented by the theory of relativity as an extension at all. The answer to this is simple and was given by Bergson long ago: because only by conceiving time as an extension, that is to say as spatial, can it be measured;[49] and measurement, as was said before, is the basis of science.

What *is* measurement? It is the assignment of numbers for the expression of homogeneous quantities as *scalars*, that is to say as collections of units which can be represented as distances proportionately marked upon a scale, lines of definite length.[50] The numbers denote those proportionate distances, those extended collections of units, the lengths of those lines, whether the scalars ever receive actual diagrammatic representation or not. Bergson certainly made a mistake when he maintained that numbers themselves are collections of units and as such are necessarily conceived spatially,[51] as Bertrand Russell has properly shown;[52] but Russell's subsequent point that even the spatial conception of collections of units merely derives from Bergson's idiosyncrasy as a visualist and is not at all a necessity of thought, is here irrelevant, because the collections of units in question, those obtained by measurement, are extended by definition, and so therefore is the succession of numbers assigned to those units.

Any measurement implies the possibility of scalar representation. If the kind of quantity measured is conceived as elementary, then its scalars are regarded as segments of an ideal line, the 'dimension' of that quantity, which together with other such ideal lines representing other elementary kinds of quantity may be regarded as forming ideal

[48] *The Leibniz-Clarke Correspondence*, Leibniz' Fifth Paper, par. 29.

[49] *Time and Free Will*, ch. 2, passim, esp, p. 116. Cf. Scott Blair, *Measurements of Mind and Matter*, p. 101.

[50] Cf. Scott Blair, op. cit., p. 63; also Duncan's *Physical Similarity and Dimensional Analysis*, 1.4 and 2.1.

[51] Op. cit., p. 75; but the definition of numbers as collections of units is as old as Pythagoras.

[52] *History of Western Philosophy*, III, xxviii.

surfaces or solids representing the synthesis of those quantities, the complex concept determined by them.[53] It is in this way that the word 'dimension', although expanded from its original application to length, breadth and height to cover such patently non-spatial quantities as time and temperature, still retains its connotations of spatiality.

The conception of time as a kind of space appears particularly appropriate because it is difficult for us to think of it otherwise than as an extension, albeit a moving one; which is why, when Minkowski stopped it moving and Weyl still wanted to think about it as a human being and not as a mathematician, he made consciousness move instead. We cannot even speak of time without spatial metaphor:[54] it is long, or short; it is an ever-rolling stream; it flows and passes; and as we get older, alas, it flies. Indeed, physics could hardly have been so successful in foisting upon us its picture of time as extended in a fourth dimension related to the three dimensions of space, had some such picture not already been natural to us.[55]

Time spatialized for the purposes of scientific measurement did not come in with relativity theory; it is as old as Galileo, when in order to express problems of motion he regarded time as measurable by a point moving along a line.[56] And we have all, now we come to think about it, been familiar with the spatial representation of time from our childhood, and of temperature too, in the fever-chart that hung on our bedrail in hospital, even though we may never have realized that our fever was thus a two-dimensional entity, and still less, which might have cheered us up somewhat, that the nurses were four-dimensional worms.

It remains to be asked why the Einstein-Minkowski world is completely static, why in fact physics, the science of nature, formally negated itself into anti-physics,[57] an unnatural geometry. Here again the answer is a simple one: because geometry is the science of space,[58] so that any theory that geometrizes time, which whatever it may be is not space, necessarily makes time spatial and motion motionless. A 'world' so constructed *must* be static.

[53] E.g. the 'colour-solid' with its three dimensions of brightness, chromaticity and saturation; v. Scott Blair, op. cit., pp. 52-54.

[54] Errol E. Harris, 'Time and Change', *Mind*, April 1957.

[55] Cf. Bergson, op. cit., p. 85.

[56] Whitrow, *The Structure and Evolution of the Universe*, p. 55.

[57] Parmenides, whose system denied motion, was called ἀφυσικός by Aristotle for that very reason, as I was pleased to find in G. R. Levy's wonderful book *The Gate of Horn*, p. 310.

[58] Whitrow, op. cit., p. 53.

Such a perversion of geometry can generate only monstrosities if embraced as true to nature. Yet that is what its sponsors appear to have demanded. 'The ideal aspiration, the ultimate aim of the theory,' Schrödinger wrote, 'is not more and not less than this: A four-dimensional continuum endowed with a certain intrinsic geometric structure, a structure that is subject to certain inherent purely geometrical laws, is to be an adequate model or picture of the "real world around us in space and time" with all that it contains and including its total behaviour, the display of all events going on in it.'[59]

A conceptual system that shall constitute adequate *knowledge* of the real world is, indeed, the ideal end not only of relativity theory but of all science; but time geometrized in a four-dimensional continuum is an inert first principle that inevitably produces a stillborn issue. To picture 'the real world around us' as motionless may have been necessary to reduce physics to a comprehensive system of interrelated differential equations, and such a system may well give more thorough knowledge of certain world-processes; but as a *picture* it is absurd. And to project such an artifact back upon living nature and declare it to be 'more objective' than nature itself is preposterous, literally preposterous, putting that before which should come after, thoughts before things and not things before thoughts.

The final truth of the matter is that the four-dimensional space-time continuum is not a physical entity at all. It is a highly abstract conceptual framework for the geometrization of relativity equations, a mere 'device for computation';[60] and after the work of E. A. Milne, who whatever else he did with time at least made this point clear, and incidentally set the scientific world-model moving again,[61] no scientist in his senses now believes it to be otherwise. Time hypostatized as a fourth dimension is the metaphysical bastard of Einsteinian physics.

4

The essential factor in relativity computations is *c*, the velocity of light in free space. This is the hub upon which the entire theory of relativity revolves. To this hub all its equations are geared. The constant value of *c* for all observers, whatever their uniform velocity relative to each other or to the light-source itself, was postulated by

[59] *Space-Time Structure*, introduction.

[60] Dr Martin Johnson, *Art and Scientific Thought*; cf. *Chambers' Encyclopaedia*, 1955 edition, art. 'Relativity': 'The object of the device is to help us in our calculations; whether its use has any metaphysical significance is a question lying outside the scope of science.'

[61] 'We superpose *motion* on geometry'—*Kinematic Relativity*, § 5.

Einstein in 1905 as the only proper explanation of the negative result of the famous Michelson-Morley experiment that in 1887 had heralded the breakdown of Newtonian physics.

This experiment, planned to discover the speed of the earth's motion in the aether, was a complete failure. Light-rays were sent from a single source at the junction of two equal rods, one of which was set lengthwise in the direction of the earth's motion around the sun, and the other at right-angles to it. The former rod was therefore travelling lengthwise at a speed of nineteen miles a second relative to the sun and at an unknown absolute speed in the aether, while the other rod had no such motion. It was naturally supposed that the light-ray sent along the rod moving lengthwise would take slightly longer to reach the end because this was receding from it, and so enable the speed of the earth's motion in the aether to be calculated. Astonishingly, it did nothing of the kind, but arrived at exactly the same time as the light-ray sent along the other rod. However often and in however different directions the experiment was repeated, the result was negative. Not only, therefore, was no absolute motion in the aether to be found, but it even appeared that the earth was not moving at all.

Only a revolutionary theory could account for this totally unexpected outcome. In 1892 Fitzgerald put forward the hypothesis, elaborated by Lorentz in 1900, that motion in the aether made bodies contract, the electromagnetic stress set up by that motion causing their particles to cohere more closely so that the lengthwise-travelling rod had shrunk exactly enough to compensate the speed of its recession from the light-ray sent along it. This, and all other hypotheses that retained the aether and the idea of absolute motion, were either disproved or shown to produce more difficulties than the one they resolved. It was Einstein who in 1905 seized the crux of it: there was no aether; the idea of a body's absolute velocity was meaningless; motion was relative to its observers; no observations made within a uniformly moving system could discover that system's motion;[62] all physical laws, whether electromagnetic or dynamical, were the same in all uniformly moving systems; for all observers in uniform relative motion the speed of light, indeed of all electromagnetic radiation, was the same. The outcome was the Special Theory of Relativity.

Now, if the observed velocity of light is neither increased by the velocity of the observer moving towards it nor diminished by the velocity of the observer moving away from it, if our measurement of

[62] The only exception is rotation (spin), which is practically absolute, not against the unsubstantial background of any aether or absolute space, but against the material background of the universe as a whole.

the speed of light reaching us from a moving system is unmodified by that system's motion relative to us the measurers, *then something else relative to us must be modified to make up the discrepancy*. This something turns out to be *the measurements which that light enables us to make*, our measurements of the moving system itself.

The greater the velocity of the observed system relative to ourselves, the more our measurements of it are distorted. All intervals upon that system, whether spatial or temporal, appear to have different values from those they would have in our own system which is relatively at rest, distances being apparently shortened and times apparently lengthened, the appropriate formulae being

$$S = s\sqrt{1 - v^2/c^2} \text{ and } T = t/\sqrt{1 - v^2/c^2},$$

where S and T are the space and time intervals as we observe them in the moving system, s and t the values those intervals would have in our own system, v the observed system's uniform velocity relative to ourselves, and c the velocity of light. These are the Lorentz transformation formulae.

It theoretically follows that if the relative speed of the observed system could possibly equal the speed of light itself, it would appear to shrink to zero. The paradox of its mass then appearing to become infinite derives from the assumption that the speed of light is an absolute maximum; not even an infinite force could impel acceleration beyond that maximum, and resistance to an infinite force requires an infinite mass.

Fortunately for sanity, all this has its converse. Since motion is relative, an observer stationed in the system so far regarded as moving and who measures intervals in *our* system from *his* viewpoint, for which it is our system that moves and not his own, finds the same modifications in *his* measurements of *our* system. For him, all *our* yardsticks look too short, all *our* clocks are slow. That is what relativity *means*: all motion is relative.

Agreement about spatial and temporal intervals between observers in relatively moving systems evidently becomes impossible, whether their observations are of each other's systems or of outside ones. Each observer's measurements of space and time are relative to his own system.

Here Minkowski had the brain-wave that made him leap out of his bath. If time-intervals are multiplied by the velocity of light and the products conceived as distances in an imaginary fourth demension, every second thus becoming 186,000 imaginary miles, then if different observers in uniform relative motion take any two events and square both the space-distance and the imaginary fourth-dimensional time-distance between those events, the difference between these squares of the space-distance and the time-distance will be the same for all of them. The square root of this difference is therefore the 'objective' interval between the two events in the imaginary absolute four-dimensional space-time continuum.

The imaginary nature of the fourth dimension is mathematically represented by applying to all expressions of it the imaginary number $\sqrt{-1}$. This symbol was originally introduced by Sir William Hamilton[63] in his quaternion calculus, an algebra in which xy does not equal yx: pouring the water on the tea and then boiling it does not, in fact, give the same result as boiling the water and then pouring it on the tea. It was used in the quaternion calculus to denote rotation through a right-angle.[64] The later use of this imaginary number has been to make any expression multiplied by it an imaginary quantity. Both uses are contained in Minkowski's fourth-dimensional coordinate $\sqrt{-1}\,ct$. Time being multiplied by the velocity of light to make it expressible as distance, ct, the multiplication of this distance by $\sqrt{-1}$ rotates it through a right-angle into an imaginary perpendicularity to all three dimensions of space so that it becomes an imaginary distance in an imaginary fourth dimension.

To come down to earth again, it should now be clear that the Special Theory of Relativity is a study in the optics of relatively moving systems. It is a study in optics because of what Ubbelohde[65] called the Primacy of Vision in our experience of the world. He proposed this primacy as 'a fundamental Law of Nature', but it hardly seems correct to call the fact that most of our knowledge of the world happens to be obtained through electromagnetic radiation of some kind a law of nature. Were it not for our fortunate possession of a couple of precious jellies sensitive to a tiny fraction of its total wave-range, the entire gamut of that radiation would have signified practically nothing. Except for the infra-red band, the rest of the range is perceptible only through instruments which convert its frequencies into sound or vision, and that only quite recently. Rather than a law, the universe of light is a privilege.

As for the central importance of the speed of light in the theory of relativity, Ubbelohde himself pointed out that time-signals can also be given by other processes, such as the radioactive emission of particles, which travel more slowly than light, thus giving different

[63] 1788-1856.

[64] For its derivation, see J. W. Dunne, *The Serial Universe*.

[65] *Time and Thermodynamics*.

relativity transformations, and he demanded that the theory be modified to take this into account.[66] However, as Einstein said, relations have to be established between different places somehow, and although theoretically it is unimportant which process is chosen, the most practical course is to choose the process of which we have most knowledge.[67]

But—and this is the point—a theory of relativity is necessary only because that process, the propagation of light, has a finite velocity. Were it infinite, the Fitzgerald-Lorentz transformations would disappear, and the theory with them. The Special Theory of Relativity was formulated to correlate inevitable optical illusions into a mathematical system. The four-dimensional space-time continuum was conceived to express that system geometrically. Within that imaginary framework relativity theory conjured by Minkowski's incantation its absolutes out of a relative world.

5

Behold, then, this patient universe unaetherized upon our table.[68] What is the diagnosis? Alas, it has anchylosis, it is *curved*.[69]

Here again, 'nothing can be done but to utter a warning that what mathematicians understand by the term "curvature" is not what the word connotes in ordinary speech. . . . Curvature (in the mathematical sense) has nothing to do with the *shape* of the space—whether it is bent or not—but is defined solely by the metric, that is to say, the way in which "distance" is defined. It is not the space that is curved,'—as though *space* could be either straight or curved!—'but the geometry of the space.'[70]

The particular geometry adopted by physicists for their 'world-models' is naturally the one that provides a framework within which they find they can give their physics its simplest geometrical expression. In short, geometry is shaped by physics; unfortunately, all too many relativity theorists have let physics be shaped by geometry.

'Curved' geometry is generally divided into two kinds, one in which parallel lines converge before infinity and the other in which they diverge; the elaboration of such systems was a nineteenth century mathematical exercise. A geometry of the first kind, the non-uniform elliptic geometry invented by Riemann in 1854, was adopted

[66] Op. cit., p. 74.

[67] *The Meaning of Relativity*, p. 27.

[68] T. S. Eliot, *The Love Song of J. Alfred Prufrock*.

[69] ἀγχύλοσ = curved.

[70] Whittaker, *From Euclid to Eddington*, I. 17.

by Einstein in 1915 for the space-time continuum of his General Theory of Relativity which incorporated gravitation by abolishing it as a distinct feature of the universe.

The gravitational mass of a body being equivalent to its inertial mass,[71] the General Theory started by regarding the motion of bodies in gravitational fields *as if* they were in inertial acceleration. Gravitational fields themselves could then be regarded as aggregates of uniformly accelerated local systems and so, appropriately geometrized, be absorbed into the space-time continuum of the Special Theory that had given absolute correlation only to inertial systems in *unaccelerated* relative motion. They distorted it considerably, for every gravitational field produced local curves in the structure;[72] but the absolute block universe filled with an infinite complex of curved world-lines that were the space-time reality of moving particles, the 'objective picture' of the world that relativity theory had set out to give, was almost complete.

It was never finished. The final aim of the theory was to incorporate also electromagnetic fields into its four-dimensional geometry. This was never achieved, and is now generally recognized to be impossible.[73]

Nevertheless, the physics that shaped that stupendous geometry is of revolutionary importance, however much this was disguised and confused by ambiguity of presentation. Mass has energy and energy has mass; or, rather, both are aspects of matter. Space and time are determined by matter; indeed, their very existence depends upon the existence of matter.[74] Within the gravitational fields of large material masses, measuring rods contract and clocks of all kinds, including atomic and physiological clocks, go slow. The greater the mass, the greater the effect.[75]

Now it must be firmly noted that these effects are *not* the Fitzgerald-Lorentz transformations, the distortions of observation produced by relative velocity of the system observed. The Fitzgerald-Lorentz transformations are *apparent*; the gravitational field transformations are *real*.

[71] The inertial mass of a body is the quantity of its resistance to acceleration; its gravitational mass is its weight; obviously, both are aspects of the quantity of its matter.

[72] Eddington, of course, preferred to suggest that the curves produced the gravitational fields and with them matter itself: *The Nature of the Physical World*, p. 127.

[73] Whitrow, op. cit., p. 96.

[74] Whittaker, op. cit., III, 46.

[75] For observations of this effect, v. Whitrow, op. cit., pp. 90-92.

The contraction of terrestrial measuring rods near bodies of greater mass than the earth is not directly measurable, because upon removal from the gravitational field that caused their contraction, they would resume their original length. But the slowing down of terrestrial *clocks* near bodies of greater mass, although their terrestrial rate would be restored upon return to earth, *would be confirmed by the real difference of time recorded.* For the Fitzgerald-Lorentz transformation this is *not* so.

This means that in so far as space-ships leave the earth's gravitational field and do not enter other gravitational fields of similar or greater intensity, their clocks will *really go faster,* although if these are observed from the earth for which they are in motion, the effect will be offset by the Fitzgerald-Lorentz transformation making them *appear* to go slower. Conversely, if space-ships enter gravitational fields of *greater intensity,* their clocks, including the physiological clocks of their occupants, *really will go slower* and not only *appear* to do so.

Why are the transformations produced by matter real and the transformations produced by velocity not real? Because velocity is relative to an observer: its *esse* is *percipi.* Matter is not relative to an observer: Berkeley notwithstanding, its existence does *not* consist in its being perceived. It would, therefore, be more correct to say not that bodies in a gravitational field behave as if they were in inertial acceleration, but that bodies in relative motion *appear* to behave as if they were in a gravitational field; but this would upset Einstein's digestion of gravitation.

Once and for all, the Fitzgerald-Lorentz contractions must be realized as what they are: optical illusions.[76] Even Eddington, after mystifying his readers for fifteen pages by talking about them as *real,*[77] later graciously admits that they are not *really real;* or, as he has to put it so as not totally to unmask his shameful abuse of the word 'real', they are *true* but not *really true,* they are *true statements about appearances.*[78] Pedlar's French like this can make anything true; it is perjury presenting itself as scientific evidence.

Is it true that partial immersion of sticks in water makes them bent? Yes, it is true but not really true. Is it true that pennies often change their shape and become elliptical? Yes, it is true, but not really true. Is there really a Man in the Moon? . . . This is indeed

[76] Whitrow's *The Structure of the Universe* makes this quite clear; v. pp. 60-61, where the word 'appear' is repeatedly used, and p. 112 where 'apparent' is in italics. Similarly, in his subsequent *The Structure and Evolution of the Universe,* v. pp. 81-83, 132.

[77] *The Nature of the Physical World,* pp. 5-19.

[78] Ibid., pp. 32-34.

ducdame, a Greek invocation to call fools into a circle,[79] the right butter-woman's rank to market[80] and the last chapter is naturally 'Science and Mysticism'.[81]

The ageless spacemen in their photon-rocket could, therefore, be dismissed as science-fiction nonsense were it not for the fact that they have been seriously contemplated by otherwise competent scientists. Eddington, of course, played about with them brilliantly.[82] But even apart from Eddington, whose popular science can no longer be taken seriously, the story has been authoritatively repeated so often that it must by now be generally believed. For example, at the International Congress of Atomic Scientists held at Chicago University in March 1958, a paper was read by a certain Dr Posin. If he was correctly reported, this gentleman stated that the inhabitants of space-ships travelling to the stars at velocities approaching the speed of light would remain young while centuries passed away on earth, and there was no indication in the press report that his fellow-scientists laughed him out of the congress hall.[83] The same fairy-tale evidently flourishes also in the other half of the world, for it has been popularized by Professors Dmitriev, Kazantsev, Landau and Rumer.[84] One can only wonder where these eminent Soviet scientists left their dialectics! On one side of the world or the other, regular newspaper readers will no doubt have seen such nonsense quite often.

It *is* nonsense. *It is nonsense because motion is relative*, so that it is one and the same thing to say that the space-ship is moving at a velocity approaching the speed of light relative to the earth as that the earth is moving at a velocity approaching the speed of light relative to the space-ship. And the very Principle of Relativity is the complete *symmetry* of observations between relatively moving systems.[85] If space-ship time appears slow to observers on earth, then earth-time will appear *equally slow* to observers in the space-ship; *but only while their relative motion lasts*. When it stops, the reciprocal illusion ceases.

[79] *As You Like It*, Act 2, Scene 5.

[80] Ibid., Act 3, Scene 2.

[81] Eddington, op. cit., Chapter XV.

[82] Ibid., pp. 38-40; also p. 135.

[83] *Il Messaggero*, Rome, 18th March 1958.

[84] *Paese Sera*, Rome, October 24, 1959, April 12, 1961 and January 11, 1962. The source for Dmitriev and Kazantsev was the Soviet magazine *Znanie Sila* ('Knowledge is Power'), and for Landau and Rumer the *Izvestia* of December 24, 1961. Profs. Landau and Rumer are joint authors of a popular book on relativity; an Italian translation of it appeared in 1961, and an English translation was expected early in 1962.

[85] Whitrow, op. cit., pp. 81-82.

The truth of the matter is, that so far from the space-travellers ageing more slowly than on earth, they would *really*[86] get old more *quickly*, because travel in free space removes them from heavy gravitational fields and so quickens all their clocks, including their physiological ones. Their only hope of stalling off old age, therefore, is to rotate interminably around a white dwarf.[87] And there we may leave them.

6

The conclusion of this discussion is that the Theory of Relativity has two things of tremendous importance to tell us about time: *qualitatively, its very existence depends upon the existence of matter; quantitatively, its real values*[88] *vary in dependent relation to material masses and its apparent values vary in dependent relation to the relative motion of those masses.*

Why, then, are we so often told nowadays that no metaphysical significance should be attached to the theory, that it is merely a matter of geometry and 'there is no metaphysics of geometry any more than there is a metaphysics of the multiplication table'?[89]

This relegation of relativity theory from the philosophical importance of its first pronouncement to the materially insignificant realm of pure geometry falls in with the general tendency of modern Western science. The process has discernible stages, which might be classified as follows, although they are often promiscuous.[90] Firstly, the incredibility of physical reality: the world is proclaimed to be utterly unlike what we experience it to be, the most astounding artifices of sophistry being employed to emphasize the contrast between the scientific world and the world we know, which thus becomes 'a world of shadows'.[91] Secondly, the inaccessibility of physical reality: science is powerless to discover the constitution of the external world and can only carry out the systematic correlation of such symbols as

[86] Yes, really really.

[87] Near the stars known as white dwarfs, the 'curvature of space-time' is terrific, because their density is over a ton per cubic inch, with a corresponding gravitational field.

[88] Each system has its own *proper time*, that of clocks stationary within it; v. Whitrow, op. cit., p. 83.

[89] Whittaker, op. cit., I, 3, as already quoted. One suspects, on the contrary, that the metaphysics of the multiplication table would be profoundly interesting, if only one had the time to go into it.

[90] The first three are found together, for example, in Eddington's *The Nature of the Physical World*.

[91] Eddington, op. cit., p. xvi.

'pointer-readings',[92] somehow produced by 'something unknown doing we know not what'.[93] Thirdly, the evaporation of physical reality: the scientific world, like the world of common experience, is an illusion, being no more than the inevitable projection of our innate mental patterns upon an unknowable background which, it is implied, is spiritual. Finally, the total irrelevance of physical reality: science becomes a collection of logically constructed linguistic or mathematical systems, and to ask what these represent is a meaningless question.[94]

We have seen this progressive mystification happening before. At their idealist rendezvous physics and philosophy meet in Platonic embrace.

It is evident, indeed, that social climate modifies scientific thought. In a society that feels itself somehow dependent upon the transcendental beliefs to which its earlier forms gave rise, scientists have an obscure duty, if not to bear active witness to those beliefs and in the last resort to do their best, as Milne did, to 'rescue God',[95] then at least to discredit and finally to disown material reality, until materialism becomes a standpoint of which we are expected to be ashamed. Those who require this would like to forget that the only purely Western standpoint *is* materialism. It must, for example, have been a very damp social climate that inspired Lancelot Hogben when he revised his *Mathematics for the Million* for its third edition and changed the word denoting his philosophical outlook from 'materialistic' to 'behaviouristic'.[96] One way or the other, of course, such processes operate everywhere, and no doubt in a society where freedom is *consciousness* of necessity,[97] conditioning is effected more crudely. We just have to make the best of it, wherever we are; but it does not help to get things clear and to see them straight, which is what philosophy ought to be about.

[92] Ibid., Chapter XII.

[93] Ibid., p. 291, where this something is aptly compared to the slithy toves gyring and gimbling in the wabe.

[94] Cf. the cogent discussion of scientific theories and reality from the Thomist viewpoint in Mascall's *Christian Theology and Natural Science*. II, 4.

[95] *Kinematic Relativity*, §268. Thomist philosophy finds no reason to be grateful, however; v. Mascall, op. cit.

[96] First page of the chapter entitled 'Statistics', p. 571 in the 2nd edition, p. 605 in the 3rd edition.

[97] Cf. Our Ford, 'My customers can have any colour they like, as long as it's black.' Or Catherine II of Russia, 'Liberty is the possibility of doing what the laws allow'; the laws of the land being, of course, equivalent with the laws of nature.

The early days of relativity theory saw the unleashing of the strangest errors. Mathematicians, mistaking geometrical abstractions for physical entities, gave their world-models a metaphysical airing amid the admiration demanded by the mystery of their differential equations; professional philosophers, happy that physics was talking their language again at last, received the revelation of hypostatized space-time with grateful respect; and as much of the theory as seeped through into the public mind gave scientific warrant to the most extravagant speculations.

Most philosophers made desperate efforts to set the space-time continuum in motion. Professor Samuel Alexander thought he could do this by declaring it to be only another name for Motion which is the absolute 'stuff of which things are made'[98] He agreed that this was rather difficult, because 'motion is in common speech merely the general name for particular motions'.[99] Nevertheless, Motion is 'a single vast entity'.[100] Space and Time may be quite empty of events or things, yet they are always full, Space being full of Time and Time being full of Space.[101] Time is the Mind of Space. . . .[102] But those who enjoy such monumental rubbish must be left to read the two volumes of *Space, Time, and Deity* for themselves.

Professor Whitehead, on the other hand, rightly censured physicists for presenting their abstractions as real entities; that, he said, was the Fallacy of Misplaced Concreteness. He then proceeded himself to display a Philosophy of Organism in which Eternal Objects constituting the Realm of Possibility achieve Concreteness by making their Ingress into the world-flux, which of itself has spatio-temporal characteristics only, so becoming Ingredients of Events which, like Leibniz' monads, mirror each other in Positive Prehensions.[103] As Dr E. L. Mascall observes, 'Most competent judges would agree that the heroic attempt of Whitehead, in his formidable work *Process and*

[98] *Space, Time and Deity*, Book I, Ch. 1; Book II, Ch. 9. For Alexander's hypostatisation of 'a fusion of geometrical coordinates', v. Martin Johnson's *Science and the Meanings of Truth*, pp. 114-115.

[99] Op. cit., Book I, Ch. 1.

[100] Idem. Jesus Christ (excuse me)! Motion of *what?*

[101] Ibid., Book I, Ch. 2.

[102] Ibid., Book III, Ch. 2. Cf. Pythagoras' saying that chronos is the psyche of the universe. But what Pythagoras meant by 'chronos' and 'psyche' could hardly have been what Alexander meant by 'time' and 'mind': chronos was cyclic, and psyche the procreative life-essence; v. Prof. R. B. Onians, *The Origins of European Thought*, II, 8.

[103] Prof. Susan Stebbing, whose healthy acidity also corroded Eddington's rot, aptly stigmatized 'Professor Whitehead's indefensible use of language'; v. Mascall's *He Who Is*, p. 158.

Reality, to interpret the world of relativity in terms of a fundamental spontaneity, avoids determinism only at the cost of unintelligibility and incoherence.'[104]

Then there was Dunne, who lacked the calibre of Alexander and Whitehead, but who achieved popularity by combining a colloquial lucidity of exposition with a sleight-of-hand that produced everybody's pet rabbits out of the Chinese boxes of an infinite regress. With Dunne, time spatialized reached its reduction to absurdity. The present moment of consciousness moving along the time-dimension necessarily takes time to do so. How quickly did the last five minutes pass? Did they take more time to pass than the previous five minutes?[105] How fast did the Time Traveller travel? At how many centuries a . . . ? There must, therefore, be another time which, again spatialized, becomes a fifth dimension perpendicular to the ordinary time-dimension. But this higher dimension too has its travelling present moment which takes time to travel and so requires another still higher dimension of time which again is regarded as an extension along which a present moment moves. . . . And so on ad infinitum.

There is nothing wrong with Dunne's argument, except its premiss. Once time is assumed to be a real extension in which things have lengths and events are situated, with a present moment moving along it into the future and leaving the past behind, or alternatively as a conveyor belt bringing events towards us, then there must be a higher time in which that motion occurs. If that higher time is similarly held to be an extension with a moving present, then a still higher time is required for *that* motion to occur in, and there is no reason why the process should ever stop. On the assumption that time is a real extension, the only alternative to the endless regress of times beyond times is to abolish motion altogether in the space-time continuum where events are situated but never happen, past and future lying qualitatively undifferentiated in imperishable coexistence; and this, however scientifically 'objective' it may be, contradicts the world we know.

Things do move and change, events do happen and cease to happen, and our knowledge of them as moving and changing and happening and ceasing to happen is real. Motion and change cannot be denied. What is wrong is not our consciousness of motion and change, but our conception of time.

Enough of philosophical history: let us now philosophize!

[104] *Christian Theology and Natural Science*. Used by permission of Longmans Green & Co., Ltd., London, and The Ronald Press Company, New York, 1956.

[105] Errol E. Harris, 'Time and Change', *Mind*, April 1957.

II

THE HERACLITEAN FIRE

'The mixer accordingly was bluntly broached, and in the best basel to boot, as to whether he was one of those lucky cocks for whom the audible-visible-gnosible-edible world existed.'
James Joyce, *Finnegans Wake.*

1

The words 'reality' and 'existence' have been so defamed that they have come to be regarded as suspicious characters and shunned accordingly. This is certainly convenient for those who wish to avoid the primary distinction between that which is and that which is not, but it does not help philosophy, if this is to be an effort to obtain clear understanding of the world and not, as Bertrand Russell once called it, a mental exercise upon simple statements to produce conclusions so paradoxical that no one will believe them.[1] We prefer to take the opposite view, that one of the tasks of philosophy is to resolve apparent paradoxes into statements so simple that every one will believe them. Dealt with straightforwardly, 'reality' and 'existence' can be quite nice words; they are anyhow more honest than such terms as 'things-in-themselves' and 'experience' that have pretended to displace them.

To announce 'The world is everything that exists' is to vie in seeming insulsity with Wittgenstein's 'The world is everything that is the case,'[2] Russell's definition of reality as 'everything you would have to mention in a complete description of the world,'[3] and the basic certitude of 'Cogito, ergo sum'. All four of these statements appear at first sight so self-evident as to be invulnerable; but upon closer inspection this is not so. Descartes' 'Cogito' makes the questionable assumption of an ego that is certainly not given by immediate consciousness,[4] as Hume pointed out; and Russell's definition

[1] 'The Philosophy of Logical Atomism', II, in *Logic and Knowledge.*
[2] *Tractatus Logico-Philosophicus,* 1,
[3] Op. cit., IV.
[4] or by psychological analysis or philosophical reflexion either: v. E.R.E., art. 'Ego'. Even Hegel reached the notion of self-consciousness only through a long travail of dialectics starting from sensation: v. *The Phenomenology of Mind.*

(the worst of the four) makes the false assumption that language is intrinsically *capable* of giving a complete description of the world, thus revealing that Russell was either quite unaware what the world is, or quite unaware what language is, or both. The two former statements, Wittgenstein's and ours, beg the question only as tautologies; they are invulnerable because 'they say nothing'.[5] But since 'the logical product of a tautology and a proposition says the same as the proposition,'[6] their latent assumptions are disclosed in their sequels.

Wittgenstein gave himself away immediately: 'The world divides into facts. Any one can either be the case or not be the case, and everything else remain the same. What is the case, the fact, is the existence of atomic facts;'[7] and so on to 'The simplest proposition, the atomic proposition, asserts the existence of an atomic fact' and 'The specification of all true elementary propositions describes the world completely,'[8] which is Russell's position as above. As Maurice Cornforth remarked in *Science versus Idealism*,[9] the method followed by Wittgenstein in subordinating the analysis of propositions to an analysis of the world makes the *Tractatus* unnecessarily hard to understand. Wittgenstein would have had us believe that he was formulating a language for the exact description of the world; actually the preposterous world that he foisted upon us was the projection of his logical syntax.

We can now, therefore, broach Wittgenstein's question and refuse to admit that the world consists of atomic facts, that its constituents are such that any one can either be the case or not and everything else remain the same, and that even an infinite aggregation of propositions could ever afford a complete description of it. On the contrary, the world is an Heraclitean[10] fire of intervolved processes, and complete description of it or indeed of any part of it is absolutely impossible, not because there is anything essentially unknowable about it, but simply because it is, literally, too rich for words. Even to attempt the irreducible task would need not the desiccated precision of the *Tractatus Logico-Philosophicus* but the protean fluidity of *Finnegans Wake*.

It really is remarkable that the very philosophers who proposed to make a 'Galilean revolution' in philosophy by the rigorous analysis of

[5] Op. cit., 4.461.

[6] 4.465.

[7] 1.2, 1.21, 2.

[8] 4.21, 4.26.

[9] 1955 edition, p. 120.

[10] This connotes the traditional view of Heraclitus, not the idealist interpretation sponsored by Diels (cf. E.R.E., art. 'Heraclitus'). It is interesting to find that Russell's *History of Western Philosophy* takes the traditional view for granted.

language displayed such crass blindness about what language is fitted to do. Take Russell's introduction to Wittgenstein's *Tractatus*, for example: 'The essential business of language is to assert or deny facts.'[11] Let us say it again, slowly this time: 'The essential business of language is to give orders.' Again, please: 'The essential business of language is to pray God.' Once more: 'The essential business of language is to ask questions.' . . . But there is no need any more to impugn Wittgenstein about this, for he did so himself in his *Philosophical Investigations*, their entire 230 close-printed pages[12] consisting in a rueful, scrupulous and painfully inconclusive broaching of the questions he had begged in the *Tractatus* twenty-seven years before. 'But how many kinds of sentence are there? . . . There are *countless* kinds: countless different kinds of use of what we call "symbols", "words", "sentences". . . . It is interesting to compare the multiplicity of the tools in language and of the ways they are used, the multiplicity of kinds of word and sentence, with what logicians have said about the structure of language. (Including the author of the *Tractatus Logico-Philosophicus*.)'[13] Thus Wittgenstein himself, sadder but wiser, twenty years on. And, not to beat about the bush, we are very soon going to beg the question ourselves.

Unless philosophy starts by begging the question, it never starts at all.[14] Solipsism, which is supposed not to beg the question, is not a start; it is a dead end. The reduction of any philosophy to solipsism is its reduction to absurdity, as consistent subjective idealists are acutely aware; hence their desperate efforts to avoid having their noses rubbed into that mess of their own making.[15] It therefore refutes itself; and, if the havoc wrought by Professor Ryle upon the Cartesian myth has not been in vain, solipsism can be shown to beg the biggest question of all.[16]

However this may be, there is one fact of paramount importance that very few philosophers have wanted to recognize, although it has been staring them in the face all the time and they have been dis-

[11] Red-here-now!

[12] '. . . *und alles, was man weiss, nicht bloss rauschen und brausen gehört hat, lässt sich in drei Worten sagen.*'—Motto of the *Tractatus*!

[13] *Philosophical Investigations*, 23.

[14] The fate of Cratylus, the philosopher who refused to beg a question and so could do no more than wag his finger, is recalled by Professor Ayer in *The Problem of Knowledge*.

[15] Perhaps the only exception was David Hume, who blandly admitted that it was a ridiculous delirium and so returned to his more sociable pastime of backgammon. (*Treatise of Human Nature*, I, iv, 7.)

[16] *The Concept of Mind*, esp. VI(1).

tressed by it to such an extent that for the past 300 years the main theme of philosophical discussion has been the epistemological problem to which it gives rise. It is, quite simply, this: that where you start, there you end. The question you beg in your beginning will, unless your logic fails, be the answer you attain in your ending. If you start in your mind you will end in your mind, and only false reasoning or divine intervention will ever get you out of it;[17] it is solipsism. Even if you start with an infinite number of minds, as Leibniz did, they will all be dreaming their separate solipsistic dreams to which only divinely pre-established harmony can give resemblance. If you start with language you will never be able to get beyond language, as Wittgenstein never could, unless to postulate everything beyond 'all-embracing logic' as 'the mystical' that could never be spoken.[18] If you start with the material world you will find no other, and Eden is lost to you for ever.

Here Thomists would protest that this is not true, because it was from the existence of the material world taken as evident that St Thomas deduced the existence of God. But that the world materially exists is only the minor premiss of the syllogism to which each of St Thomas' demonstrations can be reduced; the major premiss is the hypothetical that if finite being exists then it must be caused by absolute being. Once this is granted, you are led straight to God on any of Aquinas' five ways. But the hypothetical, in which God is already concealed, is precisely the question that is begged.[19] To start with God certainly describes the widest circle imaginable, but explicitly to *start* with God is not philosophy; as even Aquinas knew.

All philosophies, then, argue in the circle of the question, explicit or suppressed, which they beg at their outset, unless of course they go off the rails and so break down; and if they beg two questions, as Descartes' did, they run upon the rails of two separate and totally disconnected systems which only divine intervention can inexplicably coordinate. Divine intervention is not an explanation in either philosophy or science; it is a confession of philosophic or scientific impotence.

But if all philosophy is argument in a circle, and solipsism only the narrowest circle of all, why start such a barren exercise? Firstly because it is not barren and secondly because it is not an exercise. We

[17] 'We never really advance a step beyond ourselves, nor can we conceive any kind of existence but those perceptions which have appeared in that narrow compass.'—Hume, *Treatise of Human Nature,* I, ii, 6.

[18] *Tractatus Logico-Philosophicus,* 5.511, 6.522.

[19] Cf. Mascall's *He Who Is,* VI, where it is made clear that conviction by St Thomas' proofs depends upon a religious attitude to start with.

do not philosophize to exercise our wits; we philosophize because we must—to get clear the world that is before us and that requires our answer.

That an external world exists whether we are observing it or not, is with more or less reluctance generally confessed, mostly as an article of faith in a mystery transcending human apprehension, because for those who are imprisoned in their 'private perceptual worlds' the external world necessarily remains a dark nothing on the farther side of an 'unimaginable void'.[20] That we directly know the external world is a proposition which, until it was granted professional respectability after the second world-war, was maintained only by such philosophical cads as dialectical materialists and neo-scholastics; pejoratively labelled 'naive' realism, it was regarded as a superstition of common sense, which personified as a numskull was philosophically invited by Professor Broad to go out and hang itself.[21] That the external world is the *only* world we know, that the external world is, in short, *the* world, the only world that *could* be known, is the hard ground whereon, refusing to be drawn into the dark chamber of the mind[22] and there be bound by the sophistries that have deluded philosophers for the past three centuries, we take our stand.

2

The existence of highly complex material organisms alive and conscious of the world they live in, reproducing themselves in most intriguing ways and even composing symphonies and writing books on philosophy, is indeed marvellous. Matter is evidently marvellous stuff. The old view of its being mere lumps of dirt pushed about by spirits ought to be as dead as Aristotle. We have got used to the idea that given a sufficient complexity of organization matter can live. It

[20] Dr J. R. Smythies, *Analysis of Perception*, 7.1. But cf. Wittgenstein in confessional mood: 'The feeling of an unbridgeable gulf between consciousness and brain-process: how does it come about that this does not come into the considerations of our ordinary life? This idea of a difference in kind is accompanied by slight giddiness—which occurs when we are performing a piece of logical sleight of hand. . . .' (*Philosophical Investigations*, 412.)

[21] *The Mind and Its Place in Nature*, IV.

[22] *'cette cave où ils sont descendus pour se battre'* !—Descartes, *Discours de la Méthode*, VI. But it was he who led us there! . . . 'Everything is fantasy: your family, your office, your friends, your street. . . . But the most immediate truth of all is that you are pressing your head against the wall of a cell without windows and without doors.'—Kafka's *Diary*, quotation translated from an Italian review.

is time now to get used to the idea that given a sufficient development of living organisms matter can also think.

What *is* 'matter', anyhow? It is the highly abstract concept of that which in its myriad forms constitutes the world, and which it is the business of science to investigate. Beneath that investigation the concept has been transformed, the inert formless *hyle* of ancient metaphysics being revealed as an infinite complex of essentially active processes. As science penetrates those processes to deeper and deeper structural levels, qualitatively new properties of matter related by new laws emerge at each stage. Deterministic laws valid at one level are discovered to be the resultant of statistical laws appertaining in the physical substratum, and the progression may well be inexhaustible. Already the apparent finality of the probability laws of quantum theory which delighted obscurantists everywhere with the prospect of a limit permanently set to possible scientific knowledge, is giving way to the realization that quantum mechanics is itself determined by a still deeper domain.[23] But whatever the ultimate nature of matter may be, whatever the still unknown multiplicity of its ever-changing states, cannot affect the basic facts that it exists independently of all thought or consciousness, that conversely all thought and consciousness are dependent upon it, and that in all levels of being its laws, whether causal or statistical, operate whether they are under observation or not.[24] Once the essential dynamic creativeness of matter is realized, immaterial entities called minds become as superfluous to account for thinking brains as those other immaterial entities called entelechies which were once believed necessary to account for living bodies. Both are remnants of the disintegrated soul.[25]

The originally simple substance of the scholastic soul, incorporeal but incomplete without a body,[26] at least made sense with the concept of matter as it then was. But the mind, which is the soul torn by

[23] On all this, v. *Observation and Interpretation*, being the Proceedings of the Ninth Symposium (1957) of the Colston Research Society, ed. Prof. S. Körner; especially Prof. David Bohm's paper 'A Proposed Explanation of Quantum Theory in Terms of Hidden Variables at a Sub-Quantum Level' and Prof. J. P. Vigier's paper 'The Concept of Probability in the Frame of the Probabilistic and the Causal Interpretation of Quantum Mechanics'.

[24] Cf. in *Observation and Interpretation* the sharp discussion between Profs. Vigier and Rosenfeld on this very question of the objective existence of the material world and the objective validity of its laws.

[25] Cf. William James' description of the Ego: 'a cheap and nasty edition of the soul . . . as ineffectual and windy an abortion as philosophy can show.'—*The Principles of Psychology*, X.

[26] '*Anima non est persona.*'—*Institutiones Philosophiae Perennis*, II, 223.

Descartes from its essential union with the body to become a thinking ghost in a machine,[27] its prospects of immortality lost along with its simplicity,[28] is hardly intelligible.

Very few of all the philosophers who have written volumes about the mind have been able to say at all clearly what they mean by the word; some of them have never even tried. In the course of the fourteen chapters (666 pages) of *The Mind and Its Place in Nature,* for example, Professor Broad announces in the third chapter that he has been supposing that 'we know pretty well what a mind is', and goes on supposing it, whereas it is what we have been reading him to find out. We are not helped, either, when he repeatedly tells us that when he treads on a tintack a painful sensation arises in his *mind,* whereas any reader who has trodden on a tintack surely remembers that the painful sensation arose in his *foot*; which leaves one wondering whether eminent philosophers may not perhaps be a different species of being altogether. In the seventh chapter Broad states that a mind consists of a number of mental events united into a whole of characteristic structure, where the word 'structure' is, to adopt one of his own epithets, lamentably Pickwickian. Then, in the eighth chapter, he gives it up: 'I do not know how to define a mind.'

In any event Professor Broad has little comfort for those who take an exalted view of the mind's existential status: 'a human mind, taken at its face-value, is a poor sort of substance,' and he 'cannot see much reason to think that there is anything mental which is *more* substantial than finite minds, poor things as they are.'[29] His conclusion, which takes 'psychic phenomena' into account, is that 'the existence of immaterial mental substances therefore remains a mere possibility for which there appears to be no evidence whatever, normal or abnormal, a priori or empirical.'[30]

The dogma of the mind, although at last under direct attack and reduced to pitiable vagueness and ambiguity, is hard to kill. One of the more plausible arguments still supporting it relies upon the assumption that a local habitation is required for hallucinations, dreams, 'mental pictures' and such oddments as after-images. Since these presumed objects do not exist in the material world it is argued

[27] Ryle's phrase, of course; cf. Jacques Maritain's 'an angel driving a machine', quoted by Mascall in *He Who Is,* or William De Morgan's 'ghost in a corpse' (*Joseph Vance,* 1906).

[28] The primary proof of the human soul's immortality rests upon its simplicity (*Institutiones Philosophiae Perennis,* II, 212, prob. 1); conversely, the souls of animals are mortal because they are divisible (ibid., II, 184, coroll. 2-3).

[29] Op. cit., I.

[30] Ibid., XIV.

that they must exist somewhere else, namely in the 'private perceptual world' of the mind. But Hume taught us long ago that the perceptions commonly supposed to represent physical objects are similarly private, so that these too must exist in the mind. Physical objects themselves are thus reduced to mental ones—impressions or 'sense-data'. In this way that which we suppose to be material existence is shown to have no more substantiality than hallucinations or dreams, and the 'tyranny of the physical world'[31] is triumphantly overthrown.

These extraordinary conclusions are regarded as fortified by the fact that neurophysiologists have discovered that what goes on in the brain when we see something is nothing whatever like what we see, which revelation has inspired some of them to venture into the philosophical field themselves.[32] (One wonders what they expected to find in the brain. Pink rats?) On this basis Bishop Berkeley's old story can be retold in scientific tone. The new generation of philosophers has little time for the unpleasantly coloured rodents that were the favourite hallucinations of Professor Broad and his contemporaries; it prefers to sow its wild oats in the sophisticated paradise of mescaline. Elliptical pennies are scorned by those who can give themselves visions with a stroboscope or watch electro-encephalograms trace out their alpha rhythms. The latest fashion is to compare the neuro-visual system to a television set.[33] Apparently there are people who, not satisfied with spending their evenings in front of a television screen, like to imagine that they spend their whole lives looking at one. The previous generation of subjective idealists imagined a private cinema-show; the original set-up, of course, was Plato's cave. At the bottom of all such myths is a refusal of reality. Idealists are, after all, those for whom the world is not good enough; hence the common meaning of 'idealism'.

A typical representative of the neuro-physiological school of subjectivists is Dr J. R. Smythies, whose *Analysis of Perception* employs the mathematical suggestiveness of Wittgenstein's decimal notation. The mind is defined for us: '3.32. A mind is a complex composite of sense-data organized into sense-fields, together with images, thoughts, affects and perhaps a Pure Ego.' The contents of the mind form a private spatial system; indeed, the mind, ontologically distinct from the brain, must be a spatial entity; this is the only way to avoid the crippling difficulties arising from traditional Cartesian theory.[34] An

[31] Russell, 'The Philosophy of Logical Atomism', VII, in *Logic and Knowledge*.

[32] E.g., Sir W. Russell Brain, *Mind, Perception and Science*.

[33] Dr W. Grey Walter, *The Living Brain*.

[34] 3.5, 7.21, 9.

invisible and intangible physical world, which includes of course our own invisible and intangible physical bodies (*not* to be confused with the phantom bodies of our direct mental experience),[35] is a reasonable hypothesis to account for our mental experience; but its existence, although logically possible, can never be proved; and, since Smythies starts his philosophy in the private universe of his own mind, it is also logically possible that this single entity constitutes the sum total of reality.[36]. But *if* there are other minds besides Smythies', and *if* there is a physical world as well, then either all these spatial systems are self-contained universes separated by unimaginable voids, or they constitute an n-dimensional manifold with, if m is the number of human beings (but why only human?), $3m + 3$ coordinate axes, the odd 3 being for the physical world. An n-dimensional geometry is demanded to work all this out.[37] Finally, to test which of the alternatives is true, we must build a model of the brain.[38]

Considering that the largest electronic brains do not have more than ten thousand switching-units (valves) whereas the human brain contains ten thousand million of them (neurones), so that an electronic model would require a whole avenue of buildings to house it, several power stations to drive it, and a river to keep it cool, apart from the army of technicians needed to replace the thousand or so valves that would fail every minute,[39] it will be some time before Dr Smythies' theories of perception (which, he insists, are also cosmological theories) can be tested. Fortunately no such colossal enterprise is necessary; his theories collapse under the weight of their own accumulated incredibility. Fantasies like these are a waste of intellectual effort. To re-establish what Russell called the tyranny of the physical world, that is *our* programme!

We should, perhaps, be treading more softly, for we have trespassed into that strictest preserve of professional philosophy, the quagmire of theory of knowledge. It cannot be avoided, because our way goes right through it; but before invoking the protection of whatever gods look after philosophical adventurers and plunging on towards the firm ground on its farther side, let us withdraw for one long-winded paragraph into the private worlds of our minds and conjure up, if we can, a mental picture of this field of mental strife, this epistemological morass that lies before us.

35 11.

36 4.31.

37 3.181, 3.182, 7.1.

38 15.2.

39 See *Faster Than Thought*, a Symposium on Digital Computing Machines, ed. B. V. Bowden.

It is a very crowded field, rather like Hyde Park Corner on a Sunday only much worse, and even if we had time it would be hard to recognize everybody, because many of the participants have their heads completely enclosed in the black bags of their private worlds. They found themselves in this queer plight as soon as they started to think, and instead of trying to get their heads out again, they are going around demanding that black bags be accepted as the only proper philosophical headgear. A busy group of them is fitting one upon the vainly struggling sage of Koenigsberg, Immanuel Kant; others are quarrelling interminably among themselves about what they cannot see. The hubbub really is extraordinary. Nasty little devils called sense-data are rushing up and down causing much confusion everywhere. Their paternity is generally ascribed to Bertrand Russell, who haughtily wields the lever of his powerful logic, unaware that it has broken in his hands; it simply was not made for what he wanted to do with it. Dominicans, famous for their resources of persuasion, engage with polished reasons all who will listen, pointing to a large figure on the far side who must be St Thomas Aquinas, although in modern dress he looks more like G. K. Chesterton; but when we remember what lies behind their suavity they take on a sinister air, and we should do well to shun them. Cloudy entities with tattered name-tags, one of them still recognizable as Mind, are being heckled by conceptual analysts and are becoming very huffy about it. There are also a few Pure Egos, but they keep very much to themselves and most of the disputants will have nothing to do with them any more; they are just too pure to be true. Also generally ignored, but for other reasons, is the disciplined squad of dialectical materialists, chanting in unison their quotations from Marx, Engels, Lenin, Stalin[40] and Mao Tse-tung. From the platform in their midst arises the strange sight of a couple of stockinged legs waving helplessly in the air. Yes, it is Hegel, brutally captured and turned upside down; they claim it is the only way to make him talk sense. This sight of Hegel, or rather of his lower quarters, makes us wonder what has happened to Bradley who presented him to England in a haze of rhetoric, only to see him sent back to the Continent as an undesirable alien when visibility cleared. Ah, there is Bradley, and his crony Bosanquet too, dragged aside and nearly dead, but even their groans still eloquent. A more cheerful sight is David Hume, playing backgammon with opponents who, although mere mental impressions of his, seem to be beating him. Not far away is Wittgenstein, sitting sadly with his black-bagged head in his hands, fumbling with the strings. He looks worried; he cannot get out of it. Strangely enough

[40] ?—?—?—?

it has never occurred to him to wonder how he got into it.[41] And, just before we open our eyes again, behold the very high priest of subjectivism, Bishop Berkeley himself, Buckily the blodestained boyne, intoxicated with tar water, perennially slaying the perennially resurrectional Russian General of material existence![42]

Now the whole point is this: the scene we have just been trying to describe has not been taking place. Hegel was not waving his legs in the air anywhere, nor was Bradley groaning, either in this world or any other; at least, we sincerely hope not. Such pictures, which if we are visualists flicker along in the background like the running soliloquy that is the undertone of our lives if we are verbalists, and both of which can under certain conditions assume hallucinatory objectivity, are not to be located in hypothetical entities called 'minds' any more than the adventures of our dreams take place in a peculiar territory called dreamland whose frontiers we nightly cross, familiar though its several regions may become.

Dreams and mental pictures *cannot* be anywhere, for the simple reason that they are not objects. We do not dream dreams, we just dream. We do not see mental pictures, we imagine something, and *what* we imagine is *not* a mental image: think what 'imagination' means. Nor do mental pictures even appear to be anywhere. Except for after-images, and such things as hangover spots before the eyes which may, it is confessed, be mistaken for a distant flight of birds, only hallucinations *appear* to be somewhere. But they still are nowhere: that is what 'hallucination' means. We do not *see* hallucinations; we only 'see' them. We undoubtedly perceive them, but they are false perceptions. Their unreal objectivity is the self-deception of a diseased imagination.[43]

Certainly, when we imagine something, there is more than mere motion of brain-cells; we are, in fact, imagining something. Yet had we not been conscious first, we could never imagine anything. And to be conscious means to be aware of a material world.

The line is cast into the calm lake, and we sit still. Unless a breeze springs up or insects disturb us, the whole scene subsides into painted

[41] Cf. *Philosophical Investigations*, 309: 'What is your aim in philosophy?—To show the fly the way out of the fly-bottle.' But a fly isn't *born* in a fly-bottle: he must have put it there to start with! And what does a fly do when it is in a bottle? Buzz!

[42] This was intended as a contribution to the exegesis of *Finnegans Wake*, but Adaline Glasheen (*A Census of Finnegans Wake*) has arrived beforehand. Those who take years to write their book must expect its discoveries to be out-dated.

[43] Cf. Ryle, *The Concept of Mind*.

immateriality. But suddenly the line tautens, tugging at our rod, and we know at once that we have quickly to do with something really *out there,* something materially alive, and fighting.

It is not the material world that has to be proved: it is *there.* It is *there* that theory of knowledge must start if it is ever to make sense and not solipsism, if it is ever to *be* a theory of knowledge and not a theory of nescience. To ask how to get from our minds to the world is not even a serious question; it is an Alice-in-Wonderland riddle, like asking how far it is from Buddhism to Battersea. The gulf is *logically* unbridgeable, the void unimaginable. The world is *not* reached by a hazardous leap across an unbridgeable gulf. It is—look, this!

How to get from the material world to the wonderful fact that we are conscious of it, *that* is the question. But then it ceases to be a philosophical question and becomes a scientific one.

3

Creatures that make adequate responses to stimuli without ever being aware of any of them almost certainly exist; one imagines that this is how oysters carry on. Indeed, when we come to think of it, we are unconscious of many of our own responses: the normal functioning of our internal organs. A world of totally unconscious animals is quite conceivable: stimulated afferent nerves would discharge their impulses through the selective neurones into the appropriate efferents for behaviour to result, and that would be all. It is indeed wonderful that at a certain stage of evolution animals became conscious of the world they lived in, but no more than everything else is wonderful. A world of unconscious animals is conceivable, but it does not happen to be our world. Nor could it be our world. Anything more than rudimentary behaviour would need an enormous brain. It would be a world of organisms which although doubtless capable of more complex responses than those of oysters or mechanical beetles, could never achieve the terrible efficiency displayed by tigers.

The survival value of consciousness is considerable. Nervous systems sufficiently developed to translate stimuli into sensation were the necessary condition of evolutionary advance. Immobility has no use for it, but mobility had best be aware. Consciousness alone makes possible that multiform *discrimination* without which there can be no efficient action in an everchanging environment. Torn out of this context, it is a meaningless abstraction. To forget that consciousness arose in direct relation to the external world, as a means of dealing with that world, is to forget the first thing about it.

Those philosophers who, professing pure empiricism, maintain that the immediate objects of sensation are our sensations,[44] the conformity of which with the material world, if this exists, must be a matter of conjecture, ignore the empirical fact that the immediate object of sensation *is* the material world. 'Sensations' are conceptual products. We cannot recognize sensations as *ours* without self-consciousness, the realization of the self as subject distinct from the world as object, a duality already far removed from primitive immersion in the universe of being. And we cannot recognize our sensations as *sensations* until we begin, however artlessly, the philosophical exercise of introspection, which if unhealthily persisted in can lead to such strange discoveries as that my toothache (that favourite philosophers' pain) is not in my tooth, but in my 'mind'. Even when self-consciousness arises, the first word is not 'cogito' but 'sapio', with all its early connotations of tasting and smelling; not 'I think', but 'I know', verb transitive to material existence.

That philosophical analysis distinguishes our sensations from those aspects of the material world of which we are sentient, is quite legitimate; the scholastics did this long ago. Our sensations of seeing, hearing, smelling, tasting, touching, are acts whereby we have visual, auditory, olfactory, gustatory and tactual consciousness. Normally they make us conscious of the real sights, sounds, odours, flavours and contacts of material existence; but, like all acts, they can be defective or aberrant, in which event our consciousness is mistaken or hallucinated with unreal sights, sounds, odours, flavours and contacts.

Nevertheless, our acts of sensation *as such* are always real. The reality of our consciousness as distinct from the reality of that *of which* we are conscious cannot be gainsaid. We ask a mescaline addict or a crystal-gazer 'What do you see?' and he tells us. We, not seeing the shapes and colours he describes, and being able if necessary to verify his visual field by taking photographs or even analysing its light, will prefer to say he 'sees' them. But *he* says he *sees* them, and although we know *them* to be unreal, and that strictly speaking he is not *seeing* them at all, there being no such light entering his eyes, his sensations *as such* are indubitably real. Hallucinations are real sensations but false perceptions.

We say we hear a sound, but unless it is hallucinatory and hence no real sound at all, it must be a sound of *something*, even though we may not be able to decide exactly what; and certainly we should never say that the sound was in our mind, unless we were using 'sound' figuratively to mean that we were imagining it, or realized

[44] See Smythies, op. cit., 3.164, where our evidence for the existence of physical objects is said to be based on '*sensing sense-data*'!

it as hallucinatory and believed the mind to be a private world in which hallucinations have their rightful place. Now there is no such private world, there is no such place, and although hallucinations undoubtedly exist, they are not objects. We do not see hallucinations, for there is nothing to see. Our visual consciousness is being hallucinated through the misfunctioning of its neurophysiological processes.

We hear a sound. But a *sound* is not a sensation; it is our *hearing* it that is the sensation. That we can have sensations of hearing without any sound existing to be heard is easily intelligible as an incidental effect of the neurophysiological processes that make us conscious. When that happens, although our *sensation* of hearing is indisputably real, we are not really hearing at all, because there is nothing to be heard; we are just 'hearing', and what we 'hear' is unreal and so nowhere.

Our acts of sensation can, therefore, be valid or fallacious. But if they were always fallacious we should be in a hopeless state. And when these acts are metaphysically reified, when our *sensations* of seeing, hearing, smelling, tasting and touching are postulated as objects which we see, hear, smell, taste and touch, when our very consciousness is postulated to be that *of which* we are conscious, then our so-called empiricists with their 'sense-data' have gone the same way as the later medievalists with their 'species intentionales'. Both are aberrations of a decadent scholasticism.

We do not see, hear, smell, taste or touch sensations, our sensations *are* our seeing, hearing, smelling, tasting, touching, or our 'seeing', 'hearing', 'smelling', 'tasting', 'touching' as may be. Once again, they are acts, not objects. Modified by a pattern-seeking process that must presently be discussed—for consciousness, in the higher animals at least, is not mere sensation but a *conceptive* activity—their product is perception.

Now the *primary* empirical fact is that everything we perceive is perceived as objectively real[45]. This reality is not inferred; it invades us. It is the original vesture of *everything* perceived. In the beginning, *percipi est esse.* Doubt has to be *learned.*

Active experience teaches the rejection of that which illudes; the rest is reinforced by repetition to be realized as perception of a materially existing world. Dreams remain real only for the fluctuating

[45] Cf. William James, *The Principles of Psychology*, XII: 'Any object which remains uncontradicted is ipso facto believed and posited as absolute reality.' The whole chapter is important, but there is always confusion in James' thought between objective reality and subjective belief. He was, after all, an idealist.

sub-selves whose hallucinations they are; when we awake, our self reintegrates, we know them for what they were, and they vanish without harm. The terrible alternative is for hallucinations themselves to form a persistent insane whole threatening the submergence of material reality and permanent disintegration of the self.

This does *not* mean that reality is a convention. The progressive penalties for failing to learn the necessary lessons are ineptitude, misery, schizophrenia and death. It is life itself that teaches us these lessons; not contemplation, but the daily activity of living. Successfully learned, their objective result is efficient action; subjectively, it is knowledge, which from a superficial correlation of animal perceptions determined by animal needs becomes in civilized man a penetration into the essential patterns of nature, a social product, its acquisition consciously willed. But always and everywhere the criteria of knowledge are the realities of the material world. They are the arbiters, not we.[46]

The very function of consciousness makes dreams and hallucinations mere by-products. Although engaging the entire organism, it is consummated by the cerebral cortex. Sleep disengages the brain from the varied flow of continuously arriving sense-stimuli, and consciousness fails; but, like all vital organs, the brain comes to a standstill only with death. Slow rhythms discharge themselves through the nervous structure, the synapses relay their impulses to the cortex areas, and we dream. Here again, unless incoherence becomes so gross as to arouse even sleep-drugged reason to doubt, that which appears possesses objectivity by its very presence; but when the material world again invades our senses, the thin dream-pictures fade and decompose, the errant sub-selves evoked by them sink and merge into the swiftly reintegrated bodily conscious personality, and we awake. What is important is to know the difference. It is active consciousness of the physical body in the material world that holds the self together.

The conclusion is that we cannot properly be required to explain why it is that some perception has material objects, because it is the natural function of perception to present material objects. The right way of putting the question is to turn it inside out, and to ask why some perception should be delusive. Thus transposed, it hardly seems difficult to answer. Since consciousness depends upon the delicate integrity of a highly complex nervous system, there is nothing surprising in the fact that it should sometimes go wrong. Still less should such a fuss be made at finding that when the nervous system is

[46] Cf. Smythies, op. cit., 12.3, where 'the decision to call only ordinary sense experience real is a local phenomenon of Western European culture'!

intoxicated by alcohol, mescaline, santonine, hashish or marijuana, perception is deranged. There would, indeed, be call for surprise if this were not so, when mere fasting and suggestion can be sufficient to provoke such derangement.

Consciousness gives direct knowledge of the material world. None of the objections brought against this common-sense position is valid, and some of them are so puerile that it is strange how those professionally trained to think could have found them even plausible. The word 'direct' can of course be played with. The discovery that when I see something there is no skull-pent ego peering through my eyes at it, is surely not so devastating as to convince me that I never see anything in the external world at all, especially since the theory that I am, instead, seeing something in my 'mind' discloses an infinite regress.

Complaints that the act of consciousness is completed by the cortex and that neither the nerve-impulses transmitted to the brain from the sensitive periphery of the body nor the consequent excitation in the cortex itself bear any resemblance whatever to what is perceived, make one wonder how these people suppose consciouness *should* be accomplished. Are we expected to be flabbergasted because we see a tree without little tree-images travelling along our nerve-fibres and a little tree-shape impressing itself upon our brains—as though consciousness ought to be achieved by somehow looking at *that*? Are we primitive Greeks? . . .

To argue that since what we see is totally unlike what goes on in our nervous system, that which we are looking at must also be totally unlike what we see, is a mystifying non-sequitur. It is the interesting task of neurophysiologists to tell us how we perceive the world we live in. So far, unfortunately, most of them have been doing their best to convince us that we never do. That colours can be matched, symphony orchestras tuned and delicate surgical operations successfully performed is indeed wonderful. But if Sir Russell Brain is to be believed, they are so many miracles.

Another argument of the same kind is that the light or sound-waves which stimulate our senses are quite unlike what we see or hear, as though perception at a distance ought to be produced by effluvient forms; or the one popularized by Sir Arthur Eddington, that since physics has found out that material bodies are 'really' mostly 'empty space', the material bodies we see and touch are not 'really' there. All such 'scientific' arguments are fallacies; and they are used, either wittingly or unwittingly, for quite unscientific ends. They are a discredit to science, for science itself is discredited by them, the very world which is the object of scientific knowledge being proclaimed 'a world of shadows', of essentially unknowable 'things-in-themselves'.[47]

The image aroused by this subjectivist concept of 'things-in-themselves' is, unreasonably enough, a darkly *visual* one; hence the word 'shadows'. There are no corresponding images, however vague, of hearing, smell, taste or touch, because of course 'things-in-themselves' are soundless, odourless, flavourless and textureless. But it is hard to think away the *corporality* of things, which is why that muddled materialist John Locke thought he could hold on to at least some kind of external reality by distinguishing between primary and secondary qualities, the latter admittedly subjective but the former assuredly physical. The distinction is basically a Platonist one, the relative mutability of secondary qualities making them less 'real'. Anyhow, Berkeley soon showed it to be fallacious, and so made also corporality subjective.

On the assumption that if colours are not seen there are no colours, if sounds are not heard there are no sounds, if contacts are not felt there are no contacts and, finally, that if nothing is *known* there *is* nothing, everything we see, hear, feel and know the material world to be is removed from that world to the limbo of our minds; at the very most only 'things-in-themselves', empty no-things, are left. To think like this addles the brain.

Real shapes, colours, sounds and flavours do not exist because we perceive them; we perceive them when we are fitly conscious to do so. They are shapes, colours, sounds and flavours of material existence. The rainbow we see is not in our heads. It is a material aspect of material raindrops, which if we place ourselves properly we are happy to perceive. And to know how rainbows come into being does not tell use something about mind, but something about matter.

If we see a stick that, partially immersed in a stream, appears bent, this does not mean that there is a plurality of sticks, a straight one in the material world and a bent one in each of our minds. When we discover that the stick is really straight, we realize that we have not been seeing a bent stick at all, but a straight one which immersion in the stream has made to appear distorted; and we discover this not by sitting down and contemplating the stick, but by taking it out of the water and putting it in again. When we have done this several times, we have found out something; and when we know *why* partially immersed sticks appear to be bent, we have found out something more. Always, it is something about matter.

We all know that perception often has to be corrected; and it is

[47] Julien Benda wrote *La Trahison des Clercs*: it is time now for somebody to write *La Trahison des Savants*!

corrected not by introspecting our consciousness, but by *doing* something. The fact that I am very shortsighted does not mean that the blur which is all I can see when I take my glasses off has a separate blurry existence anywhere. To say I am seeing blurry sense-data is a silly way of putting it. I do *not* see blurred things: I see things blurred. It is not *what* I am seeing that is blurred, but *how* I am seeing it. It is the perceptual act that is faulty, not the world I am trying to see. I put my glasses on again, and see it better.

What we see, hear, feel and know the world to be is not something added to what it is, but partial knowledge of what it is. We can never be fully conscious of it. Were all our senses perfect, it would offer us far more; the rich olfactory consciousness of many animals is almost closed to us. All consciousness is more or less imperfect. A cow's consciousness is not a man's, and stupidity sees less in a tree than sensibility does; but it is always greater or less awareness of the same world.

What we know is limited by our senses, our instruments and our intellect. It is always partial knowledge, but the assertion that we do not really know anything unless we know everything about it, is as sleeveless as the similar argument often deployed by sense-datum theorists, that we do not really see physical objects because we only see their surfaces in perspective. Things are known in so far as what they are is known; they are unknown in so far as what they also are is not known; but they are not wholly unknowable, although we can never know them for what they wholly are. No knowledge is complete; but some knowledge is certain. Above all, the world does not lack the colours, sounds, smells, shapes and textures we perceive it to possess; it is vastly richer than them all. It is not less than we know it to be, but infinitely more.

We cannot exhaust every quibble raised by all the subjectivists who have been wrangling the question for over two centuries, nor is it necessary. Once their starting-point is realized as radically wrong, the difficulties to which it gives rise may provide philosophical puzzles, but cannot be serious problems. We know with new wonder that, despite the philosophers, the world is real after all, that long before we reached the sea its white-crested waves were sparkling in the sunlight and sounding incessantly along the beaches, and that every golden sand-grain is an infinite marvel.

4

Yet to be conscious of this boundless world to the extent that we are, immediately to respond with living awareness to the actuality that surrounds us, clear and simple though our sanity makes it seem, is achieved only by the highly delicate interaction of the most complicated nerve-structures in nature, those of our brains. Penetration into those cellular labyrinths reveals systems of bewildering intricacy. Vision alone is the integrated product of several different brain-centres, as the experience of those affected by cerebral lesions shows, or the obtaining of sight by the congenitally blind; and the other senses, too, have to be coordinated. We discover these facts with surprise, because our own infant learning has been forgotten.

Synthesis is the brain's essential activity at all levels, conscious and unconscious, perceptual and intellectual. It is biologically necessary, at the perceptual level because otherwise the nervous system could not fulfil its vital function of making the organism efficiently conscious, and at the intellectual level because redundance of information overwhelming the brain would make intelligence impossible. The key-word is pattern; or, as the scholastics told us, form. Scanning rhythms shimmering ceaselessly across the brain are broken by incoming nerve-impulses into an interplay of electric patterns that propagate their excitation through the connected circuits, an endless reverberation of patterns correlated and selected by lightning discharges and resistances of a myriad neurone-switches to produce the swift syntheses of perception, memory, imagination, thought, action.

Integration of the senses is only the beginning. Particular patterns individuated from the entire sense-field strike invisible nerve-triggers, the innate releasing mechanisms that automatically discharge instinctive behaviour in biologically appropriate situations.[48] There is still no perception of *things*. No polyp can think to itself 'Hullo, thingumbob again!' as William James supposed,[49] for it is incapable of recognizing thingumbob as a *thing* at all. A polyp has no percepts; its nervous system is selectively geared to respond only to the limited number of key-stimuli to which it must react or fail to survive, and only these demand specific awareness. Even fish are caught by a flashy contraption of feathers and steel not because they fishily misconceive it to be another fish, but because it presents the proper key-stimuli patterns; they are betrayed by their innate releasing mechanisms.[50]

Characteristic of all key-stimuli is their conspicuousness. Much of the beauty we find in nature is given by the bright colours, sym-

[48] See Prof. K. Z. Lorenz's paper on 'The Role of Gestalt Perception in Animal and Human Behaviour' in *Aspects of Form*, ed. L. L. Whyte. Cf. William James, op. cit., XXIV.

[49] Op. cit., XII.

[50] As anglers know, however, conditioned reflexes can make fishy IRMs much more selective than they originally were.

metrical designs, clear tones and intense fragrances required to force the inhibitory locks of lower animal automatisms. But the greatest stimulus of all is movement. The prey that 'shams dead' is not abandoned because it is believed to be dead, but because it no longer presents the key-stimulus. Nor, of course, is it pretending to be dead; it is frozen into immobility by a reflex that was originally a paralysis resulting from over-intensity of stimulation.[51] And movement itself receives in the sexual dance the heightened prominence of pattern.

These urgent features of the immediate environment are the first apprehensions of the animal brain, the outset of its exploration of nature to discover her significant forms. For nature is not chaos; she only appears to be so in so far as intelligence is insufficient to discover her intelligibility. Always this intelligibility is apprehended as pattern, by seizing from the protean manifold its forms of relative stability to hold them fast.[52]

Consciousness advances further when within the central nervous system computing processes of evolving complexity begin to unify perception by presenting visual key-stimuli as *constant*, thus enabling them to be recognized in varying conditions of light, distance and perspective. These are the constancy mechanisms. One of their by-products is the curious illusion that makes distant objects appear larger than by strict perspective they should do, a fact which disappoints those who have taken a photograph of a landscape including the moon. It is the shape-constancy mechanism that is the springboard for the qualitative leap to the perception of—*things*.[53]

Only a philosopher surrounded by the familiar permanence of his furniture and expecting the punctual arrival of his afternoon tea-tray could seriously maintain that the world consists of an agglomeration of particular things, atomic facts or concrete events. Out in the countryside, if he kept his senses open and his brain philosophically alert, he would surely tell a different story.[54] It is characteristic of a man-made environment that it does seem to consist of a multitude of definite *things*: they are the manifest constituents of its artificiality.

There are no *things* in nature, no problems of identity there, where

[51] Cr. James, op. cit., XXIV, where by sympathetic fallacy and inconsistently with his own theory that emotions are no more than the feeling of their physical symptoms (op. cit. XXV), he calls it a 'terror-paralysis'.

[52] It must be noted that IRMs do not cease to exist with the evolution of more complex processes; they are only submerged by them. Also higher animal and human behaviour is partly determined by IRMs. Visitors to Italy will find the sexual behaviour of many Italian males an obvious example.

[53] i.e., to Gestalt perception.

[54] Ah, there is the final testing-ground of all your philosophies—outside!

the laws of thought which apply to *things*—identity, contradiction and excluded middle—are overwhelmed. The laws of nature are transformation laws, as Hegel recognized, although he then firmly grasped the wrong end of the stick and proclaimed *those* to be the laws of thought. But it must be remarked that Russell, who had such contempt for him,[55] never got hold of the stick at all. As Russell himself admits, he never understood Hegel, and the reason was not the great man's dreadfully turbid style, but mutual incompatibility of intellect. Russell was a mathematician who persistently wanted to be a philosopher; Hegel was a philosopher who occasionally tried himself out as a mathematician. The results (we can take Russell's word for Hegel's mathematics) were unfortunate.

We perceive things as separately distinct entities by synthesis from the raw deliverings of sensation, but what we perceive are not mere *objects*; they are persistently *identifiable* objects, *things*. We recognize them as the *same things* not only in the infinite variety of their appearances but even in *their own changing*. This recognition of sameness in diversity is essentially *conception*. The leap from sensation to perception is a conceptive act; the first concepts are percepts.[56] It is we who discriminate things within the material manifold, we who establish their individuality, and it is we who thus set ourselves the problems of identity, by perceiving in nature these stabilities of our conception.

General ideas, concepts of the second level, spring from the same constancy mechanism. Once things are perceived, only one more leap is needed to perceive those things as multiple forms of distinctive general patterns, the *kinds* of things they are. This second level of abstraction is attained already by the higher animals, though only to a limited extent necessarily dependent upon the limited number of particular things they recognize, determined as this is by their biological needs supplemented by as much curiosity as is a by-product of their intelligence. Animal concepts, whether of particular things or of kinds of things, are few and irresolute. It is not until the coming of man that they can be held fast and made fertile by—words.

Shaped by simian imitativeness as conventional voice-signs for objects of practical importance to the horde, words impose their stability upon that which they denote. First as names for individual

55 See his treatment of Hegel in *History of Western Philosophy*.

56 Cf. William James, op. cit., XII: 'The function by which we thus identify a numerically distinct and permanent subject of discourse is called *conception*. . . . A single individual is as much *conceived* when he is isolated and identified . . . as is the most rarified and universally applicable quality he may possess.'

persons and things and hence for kinds of things, then as names for present actions of those things,[57] language progressively enriches the primitive world with significant features. Urged by verbal intercourse, consciousness itself expands as a social product. Above all, the voice-signs represent their objects even when these are absent from perception, evoking their images by mere utterance.[58] The pithecanthropoid advances from the litter of his cave into an operative community welded together by the common tongue that constructs its purposes.

The great leap into humanity comes when names are apprehended not as isolated signs but as instances of a latent language of universal power. In that moment of illumination the age of animal cries and gutturals is left behind, a myriad doors into intelligence await the opening, the breath-spirit of creation becomes articulate, the Word reveals itself as divine.[59]

Thenceforward man imposes the signs of his language upon all things in heaven and earth. Their names identify them, filling the universe with intelligible entities, nay with intelligent beings. The first things seen were living ones, the first things named were living persons, and in the dawn of humanity everything to which a name can be given is alive, to be swayed and propitiated by incantatory speech. The trees and streams, the very stones and spaces of the air are living beings of purposeful influence, the great mountains are gods enthroned in the massed landscape. Now even their later personifications are gone, the nymphs are fled from the streams and the dryads from the forest, the high places are empty of their rulers; but still names give our world its intelligible forms; still for us all, once they have been named for us out of the starry multitude, the Plough drives nightly its slow furrow around the Pole, Orion wielding his invisible club stands hard and bright in the December sky.

Here at last, then, is the truth in idealism. The world of our thought is the world of our language. In very truth 'the limits of my language

[57] The child repeats this development early in its second year, when it is actually taught to see things by hearing their names repeated as they are presented to its attention. Deaf-mutes have the poor surrogate of lip-movements; but 'the enormous difficulties of deaf-mute and especially of idiot instruction is principally due to the slow and painful manner in which we succeed in bringing out from the general confusion of perception single items with sufficient sharpness.'—Waitz, *Lehrbuch der Psychologie*, quoted by James, op. cit., XI.

[58] Language makes all the difference between cumbersome barter and exchange in cash or, better still, by cheque. The chance has to be taken that, like cheques, some words bounce. The reason is the same—no material backing.

[59] Cf. 'the radiant flash of intelligence and glow of joy' that accompanied this moment in the blind deaf-mute Laura Bridgman.—James, op. cit., XXII.

mean the limits' not of *the* world but 'of *my* world'.[60] The subject-predicate world of our thought *is* created for us by the Word. When the Elohim began to create the heaven and earth,[61] the earth was without form and empty; unseen were the features of its fluidity, the breath of the Elohim brooded upon the flux. And the Elohim, the deified ancestors,[62] said: Let there be Light; and there was light. The Elohim called the light Day, and the darkness they called Night; and the evening and the morning were the first day. The Elohim called the firmament Heaven, the dry land Earth, and the gathering together of the waters called they Seas; and they saw that it was good.[63]

In the beginning was the Word. Yes: but the Logos itself was born from the primordial divinity of that great Mother who, under so many titles, has her eternal being as fluid nature and fountain of living forms, as that virgin matrix divided into multiplicity by creative breath at the foundation of our world, and who in the apocalypse of the Word triumphant is beheld as a great wonder clothed with the sun and crowned with stars, mater creatoris, divine matter, the mother of all that is.[64]

5

The apotheosis of the word culminates in its transcendance of the sign to become the very vehicle of thought. Signs can do no more than denote the things they name. Speech thus restricted reflects a world that is recognized but cannot be understood. Impelled by need, man does far more than merely recognize things; he manipulates and transforms them. To do this effectively, he must know their properties and relations.

These properties and relations are first represented by vocables attached to the names of the things that display them,[65] and then, by the process of synthesis that characterizes the brain's activity at all

60 Wittgenstein, *Tractatus Logico-Philosophicus*, 5.6.

61 For this exegesis of *Genesis*, 1. i, v. E.R.E., art. 'Creation'.

62 See E.R.E., art. 'Ancestor-Worship and Cult of the Dead (Hebrew)'; also S. H. Hooke, *The Siege Perilous*, XVIII. Cf. I. *Samuel*, 28, xiii, where the witch of Endor raises 'elohim'.

63 'We often read the Bible together in its infernal or diabolical sense.'—William Blake, *The Marriage of Heaven and Hell*.

64 The Mariology is Allen W. Watts', *Myth and Ritual in Christianity*. Cf. the Great Goddess in G. R. Levy's *The Gate of Horn*.

65 Here again, this development is repeated by children, in the latter half of their second year, but *after* names and verbal indicatives have been combined by conjunctions and negatives into rudimentary sentences.

levels, perception itself is surpassed: the properties and relations of things are rationally generalized as *substantives* produced by inflexion from the original voice-signs. These substantives are not names, for no object is designated by them; they are themselves the form of the rational abstractions to which they give expression. Without them, unless by a paraphrase that describes their meaning, those abstractions can be neither communicated nor comprehended.

Beyond the sign, the object can be shown. Even if we did not know so before, *that* is what it *means*. But beyond the abstract noun there is no object. If we do not understand it, we are given other words; *they* are its meaning. In the last resort, if we do not understand them either, the meaning can perhaps be demonstrated, with much patient ingenuity on our teacher's part and much intelligence on ours, through things that exemplify it;[66] but even this tour-de-force cannot succeed unless the word is there ready to give the potential concept an intelligible form.[67]

When language transcends the sign-system that was its origin, man penetrates the frontiers of an ideal realm never before attained, the realm of rational patterns within the world perceived. The new words enable him to pass beyond the mere designation of things to the expression of rational thoughts about them, thoughts that will decide his practical activity towards those things and that will, if sufficiently correct, make that activity successful. Language becomes the instrument of human intelligence, embodying its multitudinous syntheses in a proliferation of linguistic forms.[68] In the same way as he who names the creatures of earth and the starry inhabitants of heaven announces their perceptual objectivity, so he who expresses the abstractions of his thought affirms their rational validity.

Language is the ancestral treasury wherein the concepts accumulated by the countless generations of our forebears are bequeathed to us for our use. Inevitably these concepts give our thoughts, our very consciousness, their mould. Some, we find, are quite false, having no correspondence with reality whatever; eventually, unless beneath disaster humanity falls back to primitive darkness, they will be discarded, their lingering content utterly lost. Many more are tainted with original falsity, but still serviceable if recognized as having revised values and warily handled. But it would be churlish to complain. What language could we ever have developed ourselves, with none to

[66] Cf. the charade proposed by Wittgenstein, *Philosophical Investigations*, 208.

[67] Actually, of course, we learn how to use abstract nouns by hearing or reading them used.

[68] Not only of nouns, but of all parts of speech; but it is the nouns that have given so much philosophical trouble.

teach us speech? What thoughts, then, what knowledge without this heritage? We must use it as best we can, and be grateful.

All thought depends upon language.[69] Thought *is* language for most of us.[70] Practice abbreviates it, of course, in the same way as we habitually read more rapidly than we speak; but without language, be it explicit or suppressed, abstract thought is impossible. The form of our language is the form of our thought, the limits of our language are the limits of our thought. Far from being a mysterious process self-generated in individual isolation, thought itself develops with language as a social product.

It follows that animals cannot properly be said to think at all; such 'thoughts' as they may be supposed to have must be either very simple or extremely woolly, and their animal cries and gestures no doubt sufficient to express them. Pithecanthropus himself can hardly have started with much more. Deaf-mutes, for all their inherited brain-capacity, are practically debarred from abstract thought until they learn to read. Like Wittgenstein,[71] we are quite unconvinced by William James' account[72] of the high-flown thoughts of an uneducated deaf-mute boy. Even those whose thinking is not consciously verbal have to express their thoughts in language before they can be accepted as such.

'I know what I want to say, but I can't find the words.' But you don't! You only know *what* you wanted to say when, aloud or silently, you have actually said it. Until then you only know *that* you want to say something. 'No, that's still not quite what I wanted to say; I ought to have put it differently.' Put it differently, then! It's the only way to make yourself clear, whether to yourself or to any one else, unless he or she does it for you. 'Ah yes, that's what I meant. How quick you are! . . .' Ineffable thoughts are at least shapeless; more often than not they are ineffable nonsense.

Every word expresses a concept; conversely, a concept is definable as that which is conveyed by a word. First-level concepts are abstracted directly from the material world by perception, as particular things and their recollected images; they are denoted by the verbal

[69] See Wittgenstein, op. cit., 335, 336, with his story about the Frenchman who claimed French to be the best language because its words follow the natural order of thoughts.

[70] If human testimony is to be believed, some people do think without a conscious verbal basis; but even James, despite his resistance to the thesis that all thought is verbal, declares that 'the *simple* central process is to speak when we think; to think silently involves a check in addition'.—Op. cit., III.

[71] Op. cit., 342

[72] Op. cit., IX.

signs that are their names. Higher-level concepts are abstracted from them rationally, as verbally embodied ideas. Beyond the superficial classes, properties and relations of things there are deeper ones, and beyond these again the vast patterns of universality, a whole world of interrelated meaning to be investigated and comprehended. All this knowledge is on different conceptual levels. The more general the synthesis, the higher its concept-level; the deeper the underlying patterns, the greater the abstraction of the concepts wherein they are realized. The uniform structure of language incorporates all of them.

Every word, isolated, expresses a concept, if its isolation makes sense at all; and our thoughts are expressed in language, in sets of words. Yet to pretend that thought consists of a linear procession of the concepts severally expressed by the words in which it is formulated, as though each successive word were a rigid atom of meaning unaffected by its context, would be to commit as flagrant an absurdity as any that ever leaked from a logical atomist's pen. 'I'—'I do'—'I do not'—'I do not have anything'—'I do not have anything to do'—'I do not have anything to do with it', or, 'I do not have anything to do it with'. Are the concepts 'I' and 'do' first juxtaposed in provisional affirmation before being smitten by the negative hard upon their heels? And what concept is expressed by the auxiliary? . . . How spurious these concept-strings immediately appear! If we went on thinking about the elements of our speech like this, we should soon stammer; if we went on talking about thought like this, we should be stultified.

In fact, we do not think in words at all; we think in phrases. The familiar patterns are seized or expressed in a single act. Why, it's like riding a bicycle, or walking! If we scrutinize the links of our performance hard enough to dispossess the pattern, we totter and fall again, as we did when we first began. But the simile is insufficient, for the process of thought is more than the tally of its components, howevery smoothly amalgamated into accomplishment these may be. Speech and thought are not amalgams, but compounds; and a compound is not the mere blending of the elements into which it can be analysed, nor its properties the mere summation of theirs, but new, even unpredictable. 'I do not have anything to do with it'—'I do not have anything to do it with': alter one atom's position in the organic structure, and there is another substance.

Our learning of a foreign language hardly starts until we are initiated into its syntax; before then we can do little more than cull misapprehensions in the thorny forest of its dictionary. And not until that syntax loses its stiff outlandishness and becomes a familiar habit which, donned at our pleasure, gives our thought its very form, so that *we actually think in that other language*, can we rightly be said to *know* it. The rest is easy enlargement of vocabulary to enable our thought to expand to its accustomed fullness.

The extent to which, if at all, we consciously accompany our uncommunicated perception or imagination of particular objects by inward enunciation of their names, or to which, if at all, names spoken or seen arouse images of the persons or things they signify, constitutes our psychological temperament as verbalists or visualists. Actually, of course, the opposite extremes are spanned by a gamut of individual gradations. Thought classifies, but nature confounds. Although most of us can be classified as predominantly one or the other, we are neither exclusively; it depends what we think about. But for extreme visualism, take Rimbaud: 'A noir, E blanc, I rouge, U vert, O bleu, voyelles. . . .'; and for extreme verbalism, James Joyce's description of the word 'heliotrope': 'Up tighty in the front, down again on the loose, drim and drumming on her back and a pop from her whistle';[73] or, indeed, Leopold Bloom himself.

The imagination of extreme verbalists is entirely verbal; their very dreams are monologues. Visualist prejudice, accustomed to imaginative colour and hardly suspecting the rich concordances that privately delight the verbalist, feels that this must be very dull. In return, the verbalist has little envy of what he considers to be picture-book simplicity. As opposite extremes, they are fated to mutual incomprehension.

On the first concept-level, that of perception, it is of no philosophical consequence either way, for the relation between names and things is not one of hypostatic union but of consciously separable association.[74] But beyond that level words are, as we have seen, no longer signs of the concepts they express, separable from them: *they contain them as their vehicle and embodied form*. The vehicle can be transmuted, words can be translated into other tongues, even perhaps into mathematical or logical symbolism, and the concepts themselves more or less retain their identity, modified only by the different associations of their new idiom; but without a vehicle at all they either precipitate into false images or dissolve into elusive wraiths.

Translation of anything more than names, whether we are beginners stumbling from one term to the next or expert linguists mastering whole periods at once, must be direct from one language

[73] *Finnegans Wake*, p. 223.

[74] This peculiarity of first-level concepts fully justifies their usual separate classification as percepts; but what we have been concerned to stress is the too-often ignored conceptual nature of all knowledge, whether perceptual or rational.

to the other. There is no medium of wordless ideas, although for strong visualists there may be misconceived images. We do not first extract the concept from the word or phrase we are translating and then seek the foreign word or phrase that fits it, nor without falling into error could we do so if we tried, for 'to frame an abstract idea' apart from a term that embodies it is an impossible task. Only some language or other *can* rightly frame abstract ideas. Any hiatus between original and translation is a straining vacancy.

On the second concept-level, therefore, our two psychological types are already at cross-purposes. For the verbalist, the abstract idea of a triangle is simply 'triangle': the idea *is* the word, definable if necessary by other words, and as such it presents no difficulty. The extreme visualist, on the other hand, finds the abstract idea of a triangle inconceivable, even absurd: he demands an *image* of it, which naturally enough is only forthcoming as a triangle of particular shape. To try to generalize it may make it shimmer in his 'mind's eye' from equilateral to scalene, but it is still only an overlapping of wavering particulars, never the abstract notion of a figure bounded by three straight lines.[75] In short, he can only realize second-level concepts by reducing them to first-level ones. But still not much harm is done. The visualist is aware that the particularity of his triangle is irrelevant to theories about triangles in general; he simply takes the particular as representing the general, and one particular or another as indeed the only form in which a general idea can be understood.

It is beyond the second concept-level that visualist thinking produces aberrations of serious philosophical consequence. Whether universals are platonically conceived as metaphysical entities accessible to supernatural vision or, from the very fact that they are visually unimaginable, are concluded to be objectively meaningless, the scholastic 'flatus vocis' being echoed by the American 'semantic blank', probably depends more upon the visualist philosopher's emotional adjustment towards the changing world he lives in than to his intellectual acumen.[76] One way or the other, the visualist assumption that unless a substantive designates something that exists substantially it is without existential meaning has vitiated whole volumes of philosophical argument.

Visualists find modern physics especially difficult to understand because the reality it discusses cannot be 'mentally pictured' any more, the reason being that the deep level of material processes to

[75] Cf. Berkeley's celebrated criticism of Locke.

[76] Cf. the contrast between the vigorous nominalism of Berkeley's crusading youth and the platonic yearnings of *Siris* which he wrote in his disappointed old age.

which modern physics has penetrated is in many aspects no longer expressible in terms of *things*. However deeper science may go, it now seems certain that the fundamental reality of matter, although doubtless intelligible, is not visually imaginable. Matter does not finally consist of submicroscopic replicas of macroscopic particles whirling in orbits; no mechanical model can fit the innermost activity of the Heraclitean fire.

Nevertheless, the ability to think verbally is no guarantee of immunity from philosophical error. Verbal thinking is clearly an enfranchisement, but like all enfranchisements it brings new dangers. Once disengaged from anchorage to the imaginable, thought uncontrolled can soar into an empyrean of vacuity. If the visualist's mistake is to hold abstract concepts at the concrete level, the verbalist's mistake is to forget that if they are to retain meaning they must ultimately refer to material processes.

Worst of all is the 'scientific' verbalist, who haughtily disclaims that his words are even supposed to have any material meaning. They make sense within the self-contained system of his language, and that is all. True, there are sentences that verbalize perceptual experience, but it is meaningless to ask if they correspond to material reality. The ideal language[77] would be one in which such questions simply could not *be* asked. All we can do is to compare those sentences with others and see if they are coherent. If so, they are verbally correct; if not, either they must be rejected or the entire language must be altered to accommodate them. Still less may we attribute external meaning to the concepts that correlate those basic sentences into a total linguistic system; they are mere rationalizations, convenient logical fictions, expressing not the world's nature but, in the last analysis, our own.

Thus, whilst the everyday verbalist merely tends to use words that have no real meaning, the sophisticated one denies that outside language words *can* have any meaning. On this basis science becomes a formally coherent language-system correlating the elementary 'protocol' statements that verbalize empirical data. Its only meaning lies within its own syntactical structure, its only truth in its own internal coherence. The belief that science constitutes real knowledge of a materially existent world is dismissed as 'metaphysics'. The only task left to philosophy, therefore, is linguistic analysis, until positivist paradoxes degenerate into existentialist paroxysms.

These 'physicalist' doctrines, which for all their formal originality are substantially a verbalist rehash of Kantian idealism, tacitly make over to mysticism the very world that is the object of scientific in-

[77] Carnap's 'formal mode'.

vestigation. They would cast 'metaphysics' into the opprobrium that long ago overwhelmed 'sophistry'. Left in such hands as these, it would not take long for 'philosophy' itself to sink to a similar end.

6

A supernatural origin of intellect is belied by its original concreteness. Primitive thought is entirely visual; the abstract as such is unthinkable for it; there is only metaphor. Verbal thinking evolves from the alphabet.

Yet although all primitive thinking is visual, it would be fallacious to suppose that visual thinking is necessarily primitive. Aristotle, affirming that 'the soul never thinks without an image',[78] seems to have been unaware that verbal thinking was even possible, despite the complaint made by Antisthenes the Cynic half a century earlier: 'I see a horse; horsiness I do not see.'[79] So far from visual thinking being intellectually incapable, it has accomplished the greater part of our knowledge, impregnating our entire language so that the very vocabulary in which we now think verbally is drenched in the colours of its visualist origin.

So here we are! What with false images on the one hand, baseless concepts on the other, and a language that is found under analysis to incorporate different levels of abstraction, it would seem that we ought to find it difficult to think at all. Hence the logicians' mirage of a 'perfect language' so far away from the dry bones of their algebra. Yet we do think, and with more or less care, depending upon the complexity of what we have to say, can even make our meanings clear. Perhaps it is not so surprising after all, since this is what language is constituted to do.

To know the meaning of a word is simply to know how to use it, and to think is a matter of knowing the meanings of the words appropriate to the situation that makes us think, and using them accordingly; which, from infancy on, has to be learned. But it is not always simple; if we go far, the pitfalls are many. Although the rigid classification of concepts into 'types' is one of the blind alleys of logic, the effectual hierarchy of concept-levels is confounded at our philosophical peril.

What is time? Like St Augustine, until we philosophize we know

[78] *De Anima*, quoted in E.R.E., art. 'Idea'.

[79] It was Antisthenes, too, who made the notable pronouncement that, 'the examination of terms is the beginning of education', and who wrote a lost treatise *On Words* which may well have been man's first essay in conceptual analysis.

quite well what the word means; we use it correctly every day. Diluted in daily usage, it muddles no one; it intoxicates only those who take it neat. It is only when we start to think about time-in-itself that we are gravelled by a mental cramp induced, as Wittgenstein remarked in his Blue Book, because we are trying to find a substance for a substantive. The exact place of time in the conceptual hierarchy may be difficult to determine, but it is certainly high above the level of anything substantial.

Yet the worst mistake of all is to be so amazed at the discovery that many substantives have no concrete denotation, as to conclude that what they mean has no objective existence. Legerdemain with the word 'existence' can produce the most incredible results. Honesty does not exist, and nor does justice; peace is semantically suspect; we are suddenly bereft of faith, hope and charity, health, wealth and wisdom.

Let it be clear: to deny the objective significance of abstract nouns simply because they are abstract, is to deny the validity of all that human reason has wrested from primitive darkness. Without abstractions there can be no rational knowledge, for it is by knowing universals that we comprehend particulars.

Universals are real classes, series, states, actions, properties and relations of material processes, and as such they objectively exist, whether they are known or not. Countless universals exist, countless existent universals await discovery. Their intrinsicality in nature becomes extrinsicated in the predicates of rational thought. Right conception of them constitutes knowledge; misconception of them produces logical fictions.[80] The winnowing of the logical chaff from the wheat is a practical task rather than a theoretical one; its success is proven and rewarded by material or at least intellectual control of the processes conceived.

Knowledge is the ability either to do something correctly or to give logical subjects the correct predicates, to state facts.[81] The final test is not coherence but competence. Whether we are doing or saying (thinking), our knowledge is demonstrated by its consistently successful application in practice. For all but the simplest doing, verification is a social issue.

[80] For Russell, on the contrary, tables, chairs and 'practically all the familiar objects of daily life' are logical fictions!—*The Philosophy of Logical Atomism*, VI.

[81] Cf. Ryle's 'knowing how' and 'knowing that': *The Concept of Mind*, II (3). And facts are not physical entities; they are what true statements state: v. Peter Herbst's essay on 'The Nature of Facts' in *Essays in Conceptual Analysis*, ed. Flew.

In accordance with common usage, therefore, we affirm that universals exist. But can the word 'exist' have here the same meaning as it has in the affirmation that things exist? If in one breath we say that trees exist and in another that probabilities exist, are we not revealing the word's fundamental ambiguity? If so, there would be five ways out; unfortunately, they are all remarkably unattractive.

(1) We could retain the general usage but carefully distinguish 'degrees of existence' when we philosophize. But how laboriously inconclusive this would be is shown by the history of theory of types. (2) We could retain the word 'exist' only for things and use a synonym such as 'subsist' or 'have being' for valid universals.[82] But apart from the obvious futility of such proposals, philosophy is too much a private concern already, without further accretions to its special vocabulary; what *we* want is that it be a social concern again.[83] (3) We could give each 'degree of existence' a special term, coining neologisms when the short supply of synonyms runs out. But this would only magnify the previous difficulties to absurdity in the mirage of the 'perfect language'. (4) We could deny that universals can properly be said to exist at all. But this we refuse to do, for the reasons given a couple of paragraphs ago. (5) We could deny that particular things can properly be said to exist,[84] thus leaving the word for all the rest. And this startles common sense.

If we *must* choose one of these disagreeable alternatives, we are strongly tempted to follow Russell and startle common sense. We have nothing against startling common sense if it is really necessary, although it is necessary much less often than philosophers make out, and is always a danger-signal.

But perhaps common sense need not be so startled after all. Suppose we are sitting in philosophical company around a flask of wine under an oak-tree on a summer's day and one of us murmurs drowsily: 'This oak-tree we are sitting under does not exist.' What answer can we give but a groan and tell him to go to sleep? It is plain nonsense; such a phoney-philosophical statement makes him unworthy of our company. But suppose he announces instead: 'This oak-tree we are

[82] as proposed in Russell's *Problems of Philosophy*. But this was early Russell (1912); the great man has as many periods as Picasso.

[83] Mr G. J. Warnock would protest at this. He thinks philosophy should be left to professionals: see the 'Keep Out' notice plainly posted at the end of his *English Philosophy since 1900*. The rest of us, presumably, may only subscribe to *Mind*.

[84] Russell's later doctrine: v. *Introduction to Mathematical Philosophy*, XV, XV; *The Philosophy of Logical Atomism*, V, VI; *History of Western Philosophy*, I, xviii.

sitting under exists!' What can we say to that, unless we echo Bentham[85] and declare it to be nonsense on stilts? He merits not even a groan, just a stare of astonishment; the wine has evidently gone to his head.

It makes sense to say that trees exist—there are, in fact, objects which by common consent are called trees—although it is a feeble sort of sense unless we add a complement of place; similarly, it makes sense to say that hippogriffs do not exist. Also, it makes quite good sense to say that life exists on Mars or that entropy exists or that precognition does not. But it makes no sense to say that this or that particular object exists, because if it is *there*, well, there it *is*. We may doubt what to call it, mistake its right name, even find it so new that we have to invent one; but we cannot doubt that there, whatever it may be, illusion or substance, it *is*. To affirm that it exists tells us nothing about it, and a statement that says nothing is nonsense. And if it is *not* there, then it is just not there; to add that it does not exist is fatuous. We cannot say 'That little house I saw from a distance didn't exist because when I got nearer it turned out to be a haystack.' No one ever does say such things, anyhow.

If, therefore, existence is the prerogative of valid universals, O'Brien gives Winston the right answers after all. ' "Does Big Brother exist?"—"Of course he exists. The Party exists. Big Brother is the embodiment of the Party."—"Does he exist in the same way as I exist?"—"You do not exist," said O'Brien.'[86]

Nevertheless, if we do follow Russell as he fiddles the world away for us, he too will finally lead us into the bowels of a dark mountain whence there is no emergence to daylight, the labyrinth of subjective idealism. 'But I tell you, Winston, that reality is not external. Reality exists in the human mind, and nowhere else.'[87] It is the State-philosophy of 1984, a subjective idealism in which the refinement of double-think emulates the semantic achievement of Rudolf Carnap's formal and material modes.

Nor should we gain any philosophical treasure if we followed Russell only for the short distance that seems to promise a solution to our problem. O'Brien's answer is absurd, and although there may not be much sense in saying that this or that particular tree exists, it is still good sense to say both that trees do and that entropy does. The chasm between perceptible things and intelligible universals remains. To ignore it is philosophically fatal. Nobody is likely to suppose that entropy exists in the same category as trees, but philosophers who

[85] quoted in Russell's *History*, III, xxvi.
[86] George Orwell, *Nineteen Eighty-Four*, III, 2.
[87] Idem.

substantialized motion[88] or events[89] have been acclaimed for it in universities, and even common sense blunders in its beliefs regarding mind and time.

But—and this is the point—these grave misconceptions do not arise from any ambiguity in the word 'exists'. Neither the hypostatizing philosophers nor ordinary people are misled into believing that motion or minds are substantial entities by misunderstanding what it means to say that they exist. Their errors have a deeper source. If we say that *x* exists we do *not* mean that it is substantial; we mean that it *is*. Only metaphysicians anxious to label every imaginable distinction are worried because the word 'exists' applies both to things and to their properties and relations. No such distinction is necessary; if it were, there would be a word for it already. It is unnecessary not because existential categories are always clear, but because it is not the word 'exists' that confuses us about them. If we do wish to discuss in terms of existence whether *x* is substantial or not, then we use some such phrase as 'physically existent'; but it is a roundabout way to do so. To say that minds exist is not ambiguous because of any ambiguity of 'exist', but because of the ambiguity of 'minds'.

'Existence' is *not* ambiguous. Nor, contrary to general belief, is existence a property; and since it is not a property, it cannot be possessed in greater or less degree. There are no degrees of it. Existence is not a spectrum that starts with the full-blooded being of physical bodies, fades into the pale subsistence of their qualities and evaporates through the spirituality of their relations into ultraviolet nonentity. The existence of universals is not separate from that of particulars at all; on the contrary, it is the very essence of whatever particular existence they have.

The chasm between perceptible particulars and intelligible universals is epistemological, not ontological. It is in our thought that they are separated, not in the world. Whatever exists does so absolutely, for existence is not a property, it is an act, the absolute act, the act of being.

Existence comprises everything there is—everything we see and everything we don't see, everything we know and everything we don't know—the whole infinite interplay of material processes including the wonder of consciousness, the total ever-changing splendour of the Heraclitean fire. The world is everything that exists.

[88] Alexander. [89] Whitehead.

7

This brings us to reality. Everything that exists is real and everything that is real exists. Yet 'reality' and 'existence' are not entirely synonymous. Suppose we have to comfort a drunkard in a crisis of seeing what he calls pink rats. It is no good telling him that his rats do not exist, for that only makes him angry: 'Don't tell me that nonsense! You can't see them, but I can—*there they are*!' And nonsense it is. He is not seeing non-existent rats; there is, as Dr Mascall pointed out,[90] no such species of rat as the 'rattus inexistens' which only chronic drunkards are able to see. No, what we do is to say to him gently: 'Pull yourself together, old man, they aren't *real* rats, you know! You're just *imagining them*. Yes, that's it, they're hallucinations. That's what they are *really*. . . .' And we give him a sedative.

Whereas 'existence' is abstracted from a verb, 'reality' is abstracted from an adjective. Everything which is real has reality in the same way as everybody who is healthy has health. When we know what is real we know what reality is. *Res ipsa loquitur.*

But things don't speak; it is we who have to do the talking, and words like 'real' can twist our tongues. Looked at through a philosophical microscope, the useful generality of its meaning appears so fuzzy that there is some difficulty in bringing it into focus.

The field starts to clear with the disappearance of the strange meanings reflected upon the word by transcendentalists in order to demonstrate that what we have always understood to be real is not real at all. Fortunately these artificial cloudings of its sense have not spread to obscure it altogether, though no doubt there are still a few Eddington fanatics who would argue that our table is not *really* the visibly brown and tangibly solid object we know, being *really* no more than a sparse swarm of atoms separated by relative immensities of empty space. But for the rest of us the table remains a real table, really solid and really brown.

The correct meaning or meanings of a word are decided by how it is normally used. Common sense can err, which is why we have to philosophize; but common usage is infallible.

The meaning of 'real' that emerges from common usage is given by the question '*Is it real?*' To say 'This tree is real' makes sense where 'This tree exists' is merely silly. It gives pleasant assurance of the tree's reliable solidity; we can lean against it, carve our initials on it, even climb it. Here we are tempted to make 'real' mean 'physically objective' and so divide the totality of existence into the reality of

[90] in *Christian Theology and Natural Science.*

physical objects and the unreality of their universals and all the rest. But this is barred: it flouts common usage. If the tree turned out to be a papier-maché affair put up by film people, we should no longer call it real, for all its materiality, *until we stopped calling it a tree.* Nor, even if you are a philosopher, may you tell a man that his life is not real; he would be offended. And if you tell him that death is unreal, he will think you are trying to sell him a religion.

It is not, therefore, the tree's materiality that makes us call it real, but its being essentially what we mean when we call it a tree. We mean far more than its materiality; we mean its 'treeity', that which a tree essentially is. That is its reality, its being essentially—a tree.

False percepts, for all their indisputable existence *as misperceptions*, are denied reality because they illude our consciousness. False concepts, for all their indisputable existence as *misconceptions*, are denied reality because they illude our understanding. Both miss reality by mistaking reality. Both miss reality by mistaking the essential nature of that which exists. The essential nature of that which exists *is* its reality. Reality is the essence of existence.

For confirmation, let us take the question '*What is it really?*' It is not the same to say 'This is a real tree' as to say 'This tree is real.' We could, admiring the apparition of a great tree in our dreams, exclaim 'This is a real tree!' and still sleep on; but if we declared 'This tree is real!' we should awake; the reason being that whereas *the one emphasizes form, the other emphasizes substance.* But they both come to the same thing in the end: the dream-tree is *not* a real tree, precisely because it lacks the essential substantiality of the tree that is real.

The existence of anything is *that* it is; its reality is *what* it is. We cannot think properly about anything unless to start with we know what it is, what to call it, what the right word for it is, *what its reality is*. It is the vast province of 'real' to affirm that the right word is being used. Its apparent vagueness derives not from ambiguity but from the limitless range of its application. 'Real' can qualify every noun in the language, and its adverb can qualify every verb, adjective and even preposition. It can be applied to all these words for the reason that it does not qualify what they *mean*, which would restrict it to a class, but *the words themselves*. Only indirectly does it tell us something about what they mean by affirming that, given their meaning, they are the right words. *Whenever we use the words 'real' or 'really' we are affirming, querying or denying that the words they qualify, whether explicit or implicit, are the right words.*

What are the right words? They are simply those which tell the truth.

Reality and truth are closer relations than our present mother-tongue might lead us to suppose. It was not always so. Where we now use 'real' and 'really', an older English used 'true' and 'truly', and other languages still use their equivalents. It is, for example, excellent colloquial Italian to say 'È una vera bugia!' (literally, 'It's a true lie') although one may not say 'È una bugia vera'; nevertheless, it is current usage to exclaim 'È una bugia vera e propria!'

What is truth? Affirmation cannot create it; qualification of our words by 'real' and 'really' cannot make them true. Indeed, we normally use these corroborants only when the truth is doubted.

Truth is the conformity of meaning to reality. Meaning is established by verbal usage, reality is the essential nature of that which exists. The test of truth is consistently successful application to reality in social practice.

If we call a spade a spade, we are telling the truth: a real spade is essentially a spade. But truisms like this, however necessary, do not get us very far. Although usage is an infallible guide to what words mean, common consent is certainly no guarantee that, beyond the concrete level, they are the right words. The more involved the reality to be explained, the greater the risk of mistaking the words to explain it. Either they may be the wrong words altogether, or usage may have given them wrong meanings. In either event reality is wrongly conceived and sooner or later we pay the penalty, whether in failure of action or, at the least, in confusion of thought, but always in futility. Such is the battle of rational knowledge, the battle between the meanings of our language and the realities of our world. Its real victors or victims are ourselves; its verbal victims are scattered through our dictionaries, they or their meanings being prefixed by the fatal 'Obs.'.

Perceptual knowledge, naturally enough, is on a sounder footing. We descend through doubt to converge upon certainty grasped after more or less investigation: 'That's a shilling, isn't it? Oh no, I see now, it isn't a real shilling, it's a counterfeit. But it's not even a counterfeit, really; it's just a piece of chocolate covered with silver paper. Yes, I know this silver paper is really tin. But oh dear, it's not even real chocolate—it's that nasty ersatz stuff!'

There is just time for a few questions.

'Is this a dagger which I see before me?'—'Hail, Thane of Cawdor! Nay, worthy thane, 'tis but a dagger of the mind, a false creation, proceeding from the heat-oppressed brain. In short, it isn't a dagger. Next, please! I believe the gentleman with the wooden leg wants to ask something.'

'I have a phantom limb and it often hurts me very badly. Would you say the pain was real?'—'You ought to know, sir. Pain is pain. Thank you.'

'The gentleman who has just sat down told us he had a phantom limb, that is to say a particular kind of hallucination. Do hallucinations exist?'—'Yes, they do. Had you ever had any you would not ask.'—'Are they real?'—'Real *what?*'

'*Where* do hallucinations exist, would you say?'—'Nowhere. They are hallucinations.'—'So there can be a real pain in something that isn't anywhere?'—'A pain is not a *thing*. It isn't something you're conscious *of*, but a special way of being conscious. A toothache isn't something *in* your tooth; it's how your tooth *feels*: it aches. In the case we are discussing, stimuli transmitted to the brain from the severed nerves in the stump naturally give rise to sensations of a phantom limb, and these sensations are often painful. The usual treatment is to hammer the stump until the nerve-endings become insensitive. And, by the way, a phantom limb is not a *real* hallucination, precisely because the subject is conscious of it *as an hallucination*. Technically speaking, it is a pseudo-hallucination.'

'This book of yours is supposed to be about time, isn't it? Then what has all this theory of knowledge and other philosophical backwash come into it for?'—'It's been clearing the decks and undermining time's foundations. We can close in, now.'

'Is it your opinion, Winston, that the past has real existence?'[91]—'So you're here too, O'Brien! I'm afraid your question is as Irish as your name. To ask if the past *has* existence is as absurd as to enquire if there are such things as square circles.'

'So only the present exists?'—'Nothing that did exist or will exist, but doesn't, may be spoken of as existing: the verb is in the present tense.'—'You know what St Augustine said?'—'I think I do, but let's hear it!'—'Well, this is what Augustine said: *The present has no length*.'[92]—'Ah, I was wondering when *that* was going to come up! That's how the next chapter starts. *Non più, Signor, non più di questo canto; ch'io son già rauco, e vo posarmi alquanto*.'[93]

[91] George Orwell, *Nineteen Eighty-Four*, III, 2

[92] *Confessions*, XI, xv.

[93] Ariosto, *Orlando Furioso*, XIV, 134.

III

CHRONOS DETHRONED

'And whatsoever will break on our truths, let it break! Many a house hath yet to be built. Thus spake Zarathustra.'

Nietzsche.

1

The present has no length. The point through which the future pours into the past lasts no time at all. But nothing can exist in no time at all. Therefore nothing exists.

The Platonist's escape from this old paradox is to declare that things as such[1] are neither past nor present nor future, only appearing to be so in relation to particular observers; strung out in absolute time or, after Einstein, embedded in absolute time-space, they eternally are.[2] Aristotle, on the contrary, was led by the paradox to suspect the unreality of time, for 'one would naturally suppose that what is made up of things which do not exist could have no share in reality';[3] and his conclusion that the present is carried along with material bodies in their *spatial* motion is perennially valid.

Unfortunately, the way through the symplegades of past and future has become less clear than it was in the fourth century B.C., and there is a harder headwind. Let us start by sailing into the argument that the present is relative to observers, that every 'now' is a rational soul's 'here'.

We have all heard the news that we perceive things not as they are but as they were, anything from a fraction of a second to millions of years ago, often used as an argument to conclude that we never perceive things at all.[4] Most of us, too, have seen Einstein's light-cones, if not in Einstein himself[5] then in Eddington's theatrical presentation of them,[6] a space-time hour-glass with its centre at the

[1] Modern Platonists prefer 'events'.
[2] Prof. Broad in E.R.E., art. 'Time'.
[3] Physics, IV, 10.
[4] E.g., in Smythies' *Analysis of Perception*, 3.1.
[5] *The Meaning of Relativity*, p. 36.
[6] *The Nature of the Physical World*, fig. 4.

me-here-now point of that four-dimensional worm which is myself, the lower cone representing the absolute past which is all that I-here-now can possibly see, and the upper cone the absolute future which is all that can possibly see me there, everything outside the hour-glass being, as far as I here and now am concerned, absolutely elsewhere.

There is nothing so very strange about it, except the paradoxical way it can be put. Like relativity theory, it is a matter of optics. Since ligh travels with finite velocity, it simply cannot leave more or less distant objects and reach our eyes immediately. Moreover, before we actually see those objects, another fraction of a second is needed for the machine-gun fire of impulses from our stimulated optic nerves to reach our visual brain-centres. It only seems strange to think that we never do see things immediately because normally we never do think it, and we never do because we never have to. Vision is so literally lightning-swift that its not being immediate really makes no difference, except to astronomers. If a silhouetted sentry, anxiously peering into the darkness, sees the flash of a night-sniper's rifle over six hundred yards away, he even has time to duck.

Hearing is another matter. We find nothing strange about the fact that we never hear things immediately because the relatively low speed of sound has familiar practical effects. We call, and echo answers; the nearness of a storm is judged by the time that elapses between the lightning and its thunderclap; those who make films know that in order to synchronize the sound-track they must fix it several inches before the pictures to which it corresponds; and by the time the sentry *hears* the rifle-shot, he has either been hit or not.

Only the ever-resurgent error of subjectivism can maintain that outside present consciousness there is no present existence. On the contrary, there is nothing else. So far is that which now exists from being determined by consciousness, that the farther it is away the later we are conscious of it. Whatever exists does so *now*, even though all conscious beings may perish before it can ever be seen as it now is.

Last night, looking up at the constellation of Orion, we saw Betelgeuse; the reddish light it radiated three centuries ago was entering our eyes; we *saw* it. Yet for all we really know, it no longer existed. Indeed, at a certain moment last night the whole universe outside the solar system could have been instantaneously annihilated in a single satanic act, and we never realize it as long as we live. For us, the midnight heaven would still be spangled with all its stars. Even our telescopes would not miss the first faint luminosity[7] for over four years, Sirius would not disappear for nearly a decade, and most of the great constellations would remain unmutilated for centuries.

[7] Proxima Centauri.

Before the Milky Way thinned out and finally vanished, over thirty thousand years would have to pass. A thousand generations of mankind would see a universe that no longer existed.

It is quite possible to see objects that no longer exist, if they existed far enough away, for consciousness depends upon physical processes, upon the enduring light emitted from those objects entering our eyes to give present vision of them. The absolute impossibility, rather, would be for us to have seen by physical light, at that imagined instant of apocalypse, the starry heavens rolled up like a scroll into nothingness, for that would have meant the annihilation of the past. Even omnipotence cannot perform an *impossible* miracle. That which no longer exists *cannot* be annihilated.

Consciousness cannot, therefore, let us know the present state of the nebulae, but there is no earthly reason why it should. Knowledge for its own sake is the device we have exalted from the divinely ape-like instinct of curiosity, but no such design inspired our evolutionary development. Biologically, knowledge is no more than a means to successful action. And action, too, is a means; its end is homeostasis, that is to say, happiness.

All action is towards the future. However fast our photon-rockets, they will never get us to Betelgeuse in time to catch a single one of those three hundred years of its past. As Milne told us, 'it may be shown that whatever velocity O acquires, he can never overtake events with local epochs earlier than the epoch of the event of his leaving home,'[8] although surely it needed no scientist to announce that we cannot, in fact, get anywhere before we start. On the contrary, to get to Betelgeuse will take at least three hundred years of the future, both the star's future and ours. That is the meaning of Einstein's upper light-cone. But whenever we do get there, it will then be not only our present moment and the star's present moment but the present moment of the entire universe.

This universe of present existence is not relative to any observer, for that which exists does so whether it can be observed or not. All that is relative to an observer are his observations, including his observations of clocks, a clock being definable as any process with such apparent regularity of motion or change that this can be used as a standard for the quantitative estimation either of other motion and change or of relative immobility and duration. If the successive states of a clock are serially numbered, the estimation becomes measurement. The resulting measurements are called measurements of time.

All time-measurements are relative, not merely to observers but to

[8] *Kinematic Relativity*, §54.

clocks themselves and to where those clocks materially are. That is why there are theories of relativity. The Special Theory correlates the apparently different rates of similar clocks[9] in uniform relative motion, the General Theory correlates the really different rates of similar clocks in gravitational fields of different intensity, and Kinematic Relativity correlates the really different rates of two dissimilar kinds of clock, the dynamic and the atomic. There are, too, the beginnings of a theory of biological relativity to correlate the really different rates of cellular clocks; it is, alas, the slowing down of *our* cellular clocks that produces the downhill momentum of our ageing years.

But these, however much a phenomenalist metaphysics obscures their meaning, are all theories about clocks. They have much to tell us about time and its reckoning, and with it the reckoning of the present, but they have nothing to tell us about the present itself, which is no more relative to how it is reckoned than the universe is relative to how it is observed. The present is absolute.

Let us now turn into the second tack of our argument, sailing along with this universal present, *this universe itself*, into the teeth of the windy paradox that would abolish it. However fast we go, we can never get beyond the present into the future, for like tomorrow the future is never reached. We awake into what we thought was going to be tomorrow and find it is today after all. We are still here, or nowhere. We strain our sails into the future, but it is always *now*; now, or nothing at all. We cannot get beyond what exists.

What *is* the present? Befuddled by the malleability of language, we substantivize an adjective[10] and think of the present as a point through which time passes, or we substantivize an adverb and make *now* a point that time carries along, although we should be nearer the mark if we went back to Aristotle and thought of it as carried along with the world. Its 'substratum', as Aristotle said, is always the same, being the same world, but the now itself is always different. Now is the time, but not any particular time; just *now*, whatever the time may be. Hegel, surely the greatest substantivizer ever, observed that 'a simple entity of this sort' is called a *universal*,[11] as indeed is so obvious, once we admit 'now' as a substantive, that only Russell could have wanted to maintain the contrary. He did, of course, and

[9] 'Similar' clocks are those which *in similar gravitational situations* 'keep the same time' as defined by Milne in *Kinematic Relativity*, §9; but we cannot use his word 'congruent' because he ignored the material effect of gravitation that can make them incongruent. Cf. Jeffreys, *Scientific Inference*, 8.7.

[10] etymologically a participle, verb 'praesum'.

[11] *The Phenomenology of Mind*, A.1.

reached the nadir of logical atomism when he declared that 'now' is perhaps the only word in the language that means a *particular*,[12] adding in a footnote that even this exception was open to doubt, although four years later he was prepared to add the words 'this' and 'that'.[13] It was left to Wittgenstein's sharper intellect to correct him on all three.[14]

This universal now splitting time into past and future is not an attribute of time, though Aristotle called it so.[15] Rather is it the other way about, 'time' a logical predicate of 'now', not 'now' of 'time'. Everything temporal is not present, everything present is temporal. From 'time' to 'now' the concept-level *falls*.

Wittgenstein clutched at the truth when he wrote that the function of the word 'now' is entirely different from that of a specific time-word: it is not a specification of time and *'it isn't anything like it either'*.[16] But he did not say what it *is*. Aristotle nearly said it, but for him it was easier; time as a homogeneous medium was still a metaphysical invention foreign to primitive thought,[17] and to reform academic concepts he only had to walk out of the Academy; he did not have three centuries of idealist philosophy clogging his brains to make a new concept of time hardly thinkable. Perhaps Wittgenstein did not say what that new concept must be because he *could* not; as the pioneer, he could only *show* it. And what cannot be *said* cannot, unless it is an image, be *thought*.

Words can only mean what they mean, and that is all they can say. Their meaning can be changed only by *showing in a new way* the reality to which they refer. When the meaning of words is changed, concepts are transformed. *The transformation of concepts is the changing of human mentality*. This is one of the uses of philosophy.

Misconceptions of reality can only be corrected by showing or discovering reality. Until meaning conforms to reality, reality is wrongly conceived. We cannot think rightly with wrong concepts.

If a wrong concept of time were held only by common sense, it would not matter so much; philosophy could hope, in the long run, to teach it better. Philosophers are the unacknowledged legislators of common sense; sooner or later, it conforms. Fortunately, considering the philosophical climate, it does so later rather than sooner; a few

[12] 'On the Nature of Acquaintance', II, in *Logic and Knowledge*.

[13] 'The Philosophy of Logical Atomism', II, idem.

[14] *The Brown Book*, I, 56.

[15] *Physics*, IV, 11.

[16] Loc. cit.

[17] Prof. R. B. Onians, *The Origins of European Thought*, III, 9.

of its concepts excepted, common sense still finds itself made articulate in Aristotelian scholasticism, and we can even hope that if some other philosophy becomes orthodox before common sense catches up, it may skip the past three centuries of idealism altogether, as an incoherent parenthesis.

Two common-sense concepts, however, are distinctly post-Cartesian, being the 'space' and 'time' that formed the artificial backcloth against which Newtonian science displayed its popular triumphs, and our only consolation is that the later scientific concept of 'space-time' has remained largely undigested, or matters would be even worse. In short, a wrong idea of time has become so implanted into our conceptual structure, that we cannot hold a different concept fast; no sooner is it shown to us, than it slips back before our gaze into the old illusive form again. We have to do far more than correct our inborn inclination to reify everything we conceive. What has to be overcome before we can rectify our squinted concept of time is the whole platonist metaphysics that for the past three hundred years has been making that squint so prescriptive that when we do succeed in seeing time straight for a moment, it looks queer.

The little drawings in Wittgenstein's later works,[18] shifting their appearance as we look at them, are there to show that concepts *are* changeable. They are also, no doubt, intended to show that concepts are not right or wrong, but just different ways of looking at things;[19] he was, after all, a subjective idealist to the last. But he did think harder than the rest of them, hard enough to disintegrate his own philosophical system, and he did get far enough beyond the language world of the *Tractatus* to realize that philosophy is a battle against the bewitchment of intelligence by language itself.[20]

One of the ways of breaking that bewitchment is to trace words back to their original meanings;[21] another is to find out the different implications of their foreign equivalents. By seeing through the pictorial simplicity of the original concept it is often possible to see more clearly the reality conceived; penetration into a foreign idiom often illumines the meaning of our own. We have no etymologist on board, nor can we stop our course for philological fishing, but it needs no specialist to tell us that 'present' and 'moment' are rooted in verbs of being and motion. *The reality beyond them is that which exists and changes, the universe itself.*

The present is the universal of all that is, the past is the universal of all that is no longer, and the future of all that is not yet; but as

[18] E.g., *Philosophical Investigations*, II,

[19] Ibid., I, 400, 401.

[20] Ibid., I, 109.

[21] See Onians, op. cit.

long as we go on thinking of them as extended in time, we shall never get it right. If something does not exist any longer or does not exist yet, it simply has no existence anywhere. What *is* it that exists no longer, what *is* it that exists not yet? Not the universe, surely, for the universe was, is and will be, but *particular states of its changing existence.*

Invoking Wittgenstein for the last time, we now see that St Augustine's wonderment about where his infancy had gone to when he came to boyhood, is the wonderment of the child who, watching traffic lights from a window, wants to know where the red light has gone to and where the green one comes from[22]. All Augustine's anguish at the mysterious progression of time is the anguish of the Platonist at the reality of change. And even Augustine found at last not that time produces change, but that change produces time, 'for time is made by the changes of things, as their forms vary and are changed', and 'where there is no change there is no time.'[23]

2

Our final course towards the clashing rocks of past and future bears upon the continuity of universal change that produces the continuity of time.

To speak, as we did, of particular *states* of existence implicitly belies that continuity: a moment is always a momentum, but a state is *static*. Here is an example of the most subtle enchantment that language can exert, for not only thought but consciousness itself necessarily consists in the abstraction of discontinuities from a continuous world. Until we perceive *things*, we do not perceive anything, unless we conceive *states* we cannot conceive change, without following the laws of thought we cannot think. Even philosophy cannot exorcise that basic enchantment without ceasing to be rational and ipso facto ceasing to be philosophy; but it is its duty to make us aware of it, thereby *showing* the dynamic continuity beyond. Wisdom is awareness of necessity.

Unhappily for our understanding, the 'perfect language' of mathematics hieratically binds the enchantment by *defining* change as an infinite series of static states.[24] This conception does not refer to the epistemological discontinuity of change mistakenly formulated by

[22] *The Brown Book*, I, 56.

[23] *Confessions*, XII, viii, xi.

[24] Russell, *History of Western Philosophy*, III, xxvii.

orthodox quantum theory as real,[25] but is the mathematical conception of explicitly *continuous* change, a corollary of the theory of assemblages which, based upon demonstration of the 'continuity' of numbers, defines a linear continuum as an infinite series of points.

Now this may make useful mathematics, but it makes very bad metaphysics, as Leibniz knew. It is only the lesser 'mathematical philosophers' who let their mathematics determine their philosophy. The trouble-maker is 'infinity', of course, 'sorry infinity' as Hegel called it, although his German pun is really untranslateable.[26] Every continuity is, in theory, 'infinitely' divisible; but into units, not into zeros. In theory, a line can *contain* an infinite series of points; but it cannot *be* that infinite series. An extension cannot possibly be an assemblage of points having no extension at all; the incompatibility is dimensional.

Alternatively, we can think of a line as the ideal path traced by the *motion* of a point, and we do in fact draw lines by moving the points of our pencils; but the dimensionless mathematical point is in another category altogether. By analogy, time is conceived as a continuum produced by the motion of a timeless instant, or even as a dimension along which that instant moves. All these are picturesque misconceptions.

Firstly, points and instants do not exist. They are mathematical fictions, imaginary classes with imaginary members. Even Russell, exercising his symbolic logic, demonstrated that the assumption of the existence of instants 'requires hypotheses which there is no reason to suppose true', and observed that this might be a fact of some importance in physics.[27] The misfortunes of quantum theory are, it would seem, again illuminated: false concepts of the infinitesimal cannot be applied to infinitesimal reality.

Secondly, an instant is simply a *point in time*. It cannot be conceived at all without first conceiving time as an extension, and the concept of time as a real extension is false to start with, as we saw in our first chapter.

But the sails of our argument are beating in the wind, for without using words that connote extension we cannot express the continuity of time. Spatial metaphors are all we have. The misconception forces itself upon us, inveterate in the language inherited from our visualist ancestors, predominant in our science that spatializes time in order

[25] See Prof. P. K. Feyerabend's paper 'On the Quantum-Theory of Measurement' and Dr G. Süssmann's paper 'An Analysis of Measurement' in *Observation and Interpretation*, ed. Körner.

[26] '*schlecht unendlich*': '*schlecht*' is also the obsolete form of '*schlicht*' ('simple').

[27] 'On Order in Time' in *Logic and Knowledge*.

to measure motion. To overcome this falsely extended image of the continuity of change, this metaphysical entity hypostatized from a highly abstract conception, needs effort. The last stretch towards the intermittently visible gap between past and future bends our oars.

What is the *length* of the present, how *long* does it last? The metaphor is inapplicable. The present is not a *period* of time; it is perpetual. The only possible answer in the same metaphor is that the present is as broad as the universe; incessantly changing, it always lasts. If we are then asked how long it lasts *without* changing, the only polite answer to the obstinate petitio principii is that it doesn't: the present is always changing.

But this is the *real* present. What about the present of our conscious experience, 'the short duration of which we are immediately and incessantly sensible',[28] the *specious* present? How long does *that* last?

The question is a treacherous one. Were it intended only to ask what frequencies of intermittent stimuli produce continuous sensations, it could be given admissible answers, which according to sense and sensibility would vary up to a maximum of about one tenth of a second. Such brief persistence of sensory excitation is the familiar secret of cinematography. But these indiscernible fractions are certainly not the 'specious present' that a mistaken metaphysics foisted upon an immature psychology. What the specious-present theorists maintain, James' 'saddleback' rejected, is that each instant of consciousness makes us directly aware of a short period of the immediate past terminating 'in time' as near to the point-instant of the 'objective' present as the transmission of light, sound and nerve-impulses will allow; or, rather, since all consciousness is consciousness of change, with duration as relative to it, even if that change is only our own breathing, that each instant of consciousness delivers an immediate whole of past change or duration 'varying from a few seconds to probably not more than a minute'.[29]

The theory is plausible, but false, and like all false theories it can be demonstrated to lead to paradoxical conclusions: firstly that the longer we look the shorter the change or duration we see as a whole, and secondly that any change or duration we see as a whole must be completely past before we even *begin* to see it.[30] The theory is false

[28] James, *The Principles of Psychology*, XV.

[29] Idem.

[30] For discussion of Prof. C. D. Broad's demonstration of these paradoxes, v. J. D. Mabbott, 'Our Direct Experience of Time', *Mind*, April 1951; C. W. K. Mundle, 'How Specious is the Specious Present?', *Mind*, January 1954; Mabbott, 'The Specious Present', *Mind*, July 1955.

because it depends upon the conception of consciousness as a series of *instantaneous acts*, and consciousness is no more a series of instantaneous acts than time is a series of timeless instants, space a manifold of dimensionless points, or change a succession of static states. *The specious present theory is specious because consciousness is continuous.* Much of its specious plausibility derives from our consciousness of *spatial extension*.

What is the length of the past? This metaphorical question means to ask, we suppose, what the 'age of the universe' is, that is to say, the conjectured measure of the entire continuity of past universal change as inferred by the temerarious extrapolation of *present* physical laws back to a hypothetical state of material existence beyond which those laws become *definitely* inapplicable, thus constituting the utmost possible limit of scientific thought.[31] What, then, is the age of the present universe? Whatever the various answers, the thousands of millions of years that overwhelm our imagination are not measurements of a past extension of *time*; they are speculative measurements of the total theorizable continuity of past cosmic *change*.

How long does a million years last, or a century, or a second? The answers are as tautologous as the 'eternal truth' that there are thirty-six inches in a yard, with the essential difference that whereas a yard is a spatial extension measured by a spatially extended yardstick, a minute is not a temporal extension measured by a temporarily extended standard minute, because no such extension exists to be measured. Our minute is the regulated movement of a hand from mark to mark upon a dial. Real extension, spatial extension, is measured by spatially extended meters; but the unreal extension of time can only be measured by clocks, that is to say by meters having standard *motion or change*. Clocks tell us their times, but they cannot be tallied wth temporal extension, for there is no such thing; they can only be tallied with other clocks.

Let us get it straight: *clocks do not really measure time at all.* What they measure is motion and change, or relative immobility and duration, *in terms of their own motion and change*. The result is called time; the regular progress of their motion *is* time. The moving hands of our timepieces, miniature mimics of chronos's archetypal circle,[32] turn through the successive spatial positions that number our arbitrary divisions of the earth-rotating days; the great revolving clocks

[31] Cf. Whitrow, *The Structure and Evolution of the Universe*, VIII, passim, esp. his reference to Milne's 'curious belief that the world originated at a point-source', also criticized in Mascall's *Christian Theology and Natural Science*.

[32] Onians, op. cit., II, 8.

of the universe turn through the successive spatial positions that determine the returning stages of the cyclic years.

While the earth remaineth, seedtime and harvest, and cold and heat, and summer and winter, and day and night shall not cease.[33] Their ceaselessness is a continual extinction and new creation. The rolling of the solid globe along its solar orbit makes fresh nights to wash away the stains of expired days and dissolves them into bright new mornings, gives due seasons for the sowing of the seeds left by a vanished fruit and for the gathering of the growth born of their corruption. The glory of past summers has gone with the tears of our youth; the sunlight that rejoiced us then was poured upon a world that simply is no longer *there*, that since then has passed through immense spatial vacancies and left no trace behind; we wept when the earth was somewhere else altogether. It was all long ago.

Older and wiser, *we* have remained, we are still 'ourselves'; that is to say, our existence has so far displayed sufficient continuity of form and behaviour to be called by the same name as it was then. And, of course, unlike mere things that have no other title to identity, we remember. We shall not, therefore, say with the sophists that Coriscus in the Lyceum is a different being from Coriscus in the market-place.[34] But we have surely grown older. Like everything else, we have been changing, and shall continue to do so until like Coriscus (whoever he may have been) we cease to be ourselves any longer, and become something else. Even though no Aristotle casually grant us verbal immortality, if we nourish good grass for unborn spiders to spin their webs in, it is enough[35].

What, then, is time? Its reality is so highly abstract that were we not aware of visualist inability to think without an image, it would be astonishing that it should ever have been conceived as concrete. It is a quantity of change or relative permanence. This quantity is measured by regarding certain continuous motions or serial changes as standard. As measurable quantity, time has an ideally extended scalar. The infinitely continuous change of the material universe makes that scalar ideally infinite. It becomes an imaginary dimension. As such it is divisible into smaller and smaller imaginarily extended units, and ultimately, if our conception of continuity allows, into extensionless zeros. Since it is in temporal units that we measure change and duration, it follows that the smaller the unit the less

[33] *Genesis*, 8, xxii.

[34] Aristotle, *Physics*, IV, 11.

[35] 'I bequeathe myself to the dirt, to grow from the grass I love. If you want me again, look for me under your boot-soles!'—Walt Whitman.

change or duration it measures, until the ideal zero measures no change or duration at all.

All this is certainly useful, but unless we realize what we are doing we shall utterly mistake reality. Motion and immobility, change and duration, must needs be measured, and only through the concept of time can this be done. Where we go wrong is in imagining the quantity we measure as *really extended*. The reduction of the abstractness of time to the level of physical reality is a visualist misconception; to regard a highly abstract quantity as physically extended makes philosophical nonsense.

In physical reality there *is* no such extension as time; there is just the ever-changing spatially extended world. *The world is not extended in time so that in a fraction of time there is only a fraction of world and in no time there is no world at all. Incessantly changing, the world is wholly present all the time.*

3

There is an argument that time must fundamentally be more than clock-times because we can conceive the whole universe, including our clocks and ourselves, going faster and faster until everything would take far less time than it does now. We should never realize the difference, for all our time-measurements would remain exactly the same, but their real values would be immeasurably reduced.[36] 'All motions may be accelerated or retarded, but the true or equable flowing of absolute time is liable to no change.'[37]

A similar argument can be applied to space. If the universe were to shrink until all its components and the distances between them were diminished a millionfold, everything would seem to be just the same size and as far apart as it is now. There would be no possible means of telling the reduction. No galaxies hitherto beyond vision would come into the range of the Mount Palomar telescope, for the cosmic shrinkage we are imagining is not that contraction of the universe theorized by cosmologists as the opposite process to its present hypothetical expansion which concerns only intergalactic space and has nothing to do with the size of things. It involves our very units of measurement and the speed of light itself. It is the vision of Julian of Norwich, the hazel-nut in her hand, which 'methought it might suddenly have fallen to naught for littleness'.

Now Julian's vision is valid only against the infinitude of her world's Creator, or at least against the background of Newtonian

[36] Henri Bergson, *Time and Free Will*, pp. 193-194.

[37] Sir Isaac Newton.

absolute space. Without such an absolute it is a mere trick of thought. Size is relative, and the universe is everything there is, not contained in any absolute space but its spaces determined by its material existence. To say that *everything* becomes small removes every standard that could give such smallness meaning: it postulates a relation without relata, an absurdity.

The time-argument is complicated by spin. Rotation is practically absolute, not indeed against the metaphysical background of Newtonian space but against the physical background of the universe as a whole.[38] The accelerated rotation and revolution of celestial bodies would have material consequences. If the earth were to start spinning twice as fast as it does now we should feel the centrifugal difference, even though we and everything else had been similarly accelerated. And times *would* be changed: there would be more days to the year, not because our days had shortened—they would still have their twenty-four hours and pass as slowly or quickly as they always did—but because of the widening of the terrestrial orbit. In the unlikely event of our remaining alive we might, by comparing our observations with pre-catastrophe data, be able to formulate the hypothesis that the entire celestial works were somehow revolving at twice their former rate and hence that the new times were really half the old times.

But this still gives no warrant to the doctrine of absolute time. Rates of motion are not measured by 'the true or equable progress of absolute time' but by the standard motions of clocks. There are no absolute rates of motion. The new rates are twice the old ones only if they are reckoned against the phantoms of the old clocks; otherwise they are exactly the same as they were before. Similarly, the new times are half the old times only if the new clocks are rated against those phantoms. Everything takes half the time it took before, but the shorter time it takes is illegitimately measured by standards that no longer exist. It does not take so much less *real* time, for there is no time to which that hall-mark exclusively belongs; there are just times. As the higher conceptual products of relative motions, they are related not as different quantities of any absolute time but simply by their own ratios.

Bergson cheated by supposing that universal acceleration could occur leaving us its observers exempt, thus producing the phantasm of his 'duration'. Certainly, if everything went twice as fast it would be going twice as fast, and it needs no Bergsonian intuition to tell us that if we and our physiological clocks alone remained unmodified, we should soon realize that something was wrong. We should, in

[38] Whitrow, *The Evolution and Structure of the Universe*, 3.

fact, feel ourselves bewitched into sluggishness, in the same way as if we remained 'the same size' in a shrinking Alice-in-Wonderland world we should find ourselves becoming larger and larger.

What Bergson did was to hypostatize an elusive 'mental synthesis'[39] too unstable to be a genuine concept, the 'quality' of time intuitively rather than intelligibly abstracted from its 'quantity', and exalt this metaphysical abstraction as the ultimate 'concrete'[40] reality. Since 'it is extraordinarily difficult to think of duration in its original purity,'[41] it must be 'felt' by introspection of the deeper ego wherein it resides and which indeed creates it.[42] The spurious concept of quantitative or clock-time is only duration's 'shadow', projected by 'an act of the mind' along with the equally spurious concept of extended space to transform the 'heterogenous multiplicity of conscious states' into a vision of the external world.[43]

Like all subjective idealists, Bergson held his 'mind', with however peculiar reliance upon its darker recesses, to be the arbiter of reality, if not indeed the world-egg from which it all hatched. His shifting dualism, anyhow 'transcended', is mere confusion. For Newton, 'duration' was the 'other name' for absolute time; for Bergson it was that same absolute desperately rescued from its demolished outpost in the physical universe and obscurely reinstated in a spiritual stronghold. Bergson too, it appears, hankered in secret after absolutes; he too was a Platonist at heart.

Then, with Kinematic Relativity, the absolute made a surprising counter-attack to reconquer its lost universe. For Milne, atomic time was *absolute*.[44] By applying it to the expansion of the universe boldly extrapolated backwards into the past, he reached its end or rather its beginning in a 'point-singularity' of divine creation.[45]

His fundamental data are an observer's awareness of 'the passage of time' and 'the separating cloud of particles',[46] but he soon admits[47] that this awareness of the passage of time is not any consciousness of something called 'time' and of its 'passing', but is simply the observer's ability to say which of two events 'in his own consciousness' is earlier or later. Since by hypothesis the observer observes nothing else, these events can only be his observations of the separating cloud of particles. 'The passage of time,' therefore, is simply the observer's conceptual abstraction from their *changing spatial relations*.

It is difficult to discuss the aprioristic speculation of Kinematic

[39] *Time and Free Will*, p. 120.
[40] Ibid., pp. 155, 239.
[41] Ibid., p. 106.
[42] Ibid., p. 137.
[43] Ibid., pp. 93ff.
[44] *Kinematic Relativity*, §§135, 253.
[45] Ibid., §248.
[46] Ibid., §2.
[47] Ibid., §7.

Relativity, especially when we are told that the theory is akin to the construction of an abstract geometry and that the only relevant test of its truth is its logical self-consistency;[48] but even so, the designation of atomic time as 'absolute' is plainly gratuitous, its only recommendation to that title being the simplicity of its theoretical formulation.[49] No competent admirer[50] of the theory has cared to repeat the word, and all we can imply from its use is that Milne coveted an absolute, for the same private motives that made him consciously pleased to come back to the Platonic doctrine of Ideas.[51]

Briefly, different material processes have such fundamentally different rates that in the long run they definitely do not *keep the same time*. It has been discovered that there are two *universal* processes with such fundamentally different rates, atomic processes and gravitationally dynamic processes. Both can be regarded as clocks, but eventually an atomic clock will be found to be ahead of a dynamic one. Unless, therefore, it is to be assumed that in the long run either dynamic processes really become slower or atomic processes really become faster, they must be regarded as constituting two different time-scales. Since there is no reason to suppose one universal process to be a better time-keeper than the other, the latter course is correctly preferred. Atomic processes and dynamic processes constitute *dissimilar clocks*. Neither is keeping the *right* time, because for natural clocks there is no right time; there are just times. The cosmological correlation of these two different universal times is the only genuine achievement of Kinematic Relativity. So far from justifying any time as absolute, the plurality of universal time-scales shows all times to be conceptual abstractions from different material processes.

All duration, therefore, is relative to the standard rates of change that are our clocks, and those who seek for absolutes must look elsewhere. Duration lost, they may find one in what is called the *trend* of time; that is to say, not in *quantity* of change but in the essential *seriality* of change. The serial progression of universal change produces the so-called *trend of time* popularized by Eddington as 'time's arrow'.[52]

One aspect of progressive universal change is the expansion of the universe, upon which Milne based his Kinematic Relativity. Another is the increase of universal entropy. Both are essentially *spatial*.

By the Second Law of Thermodynamics, if a closed system has its

[48] Ibid., § 5.
[49] Ibid., §§ 31, 253.
[50] E.g., Drs Whitrow and Martin Johnson.
[51] Ibid., § 5.
[52] *The Nature of the Physical World*, p. 69.

heat unevenly distributed, then any two of its total states are infallibly distinguishable as respectively earlier or later by a physical difference, the greater or lesser *spatial uniformity* of their heat-distribution, the state of greater uniformity being necessarily later; hence the 'arrow'.

Heat flows by conduction, convection or radiation from hotter regions to cooler ones. The process naturally results in a gradual levelling of temperature throughout a closed system. Since it is difference of temperature that makes heat convertible into other forms of material activity, it follows that the less differences of temperature there are within the system, the less its heat can be dynamically operative. An ever-increasing proportion of the total thermal energy of the system therefore subsides into irrecoverable inertiality. Its measure is called entropy. The progressive increase of universal entropy results from the inertial trend of universal matter towards states of greater stability. It is a trend of material motion. The trend of time is the trend of universal change.

Eddington, of course, did his best to make entropy mysterious: it is 'subjective', a statistician's 'mind-spinning', there must be 'something as yet ungrasped' behind it, some 'mystic interpretation'.[53] This is flim-flam stuff. Heat is an effect of material motion, and entropy the inertial quantity that results from the increasing evenness of its spatial distribution. There is nothing mystic about it, or subjective either. It is as matter-of-fact and objective as a death-rate or, well, a batting-average,[54] existing whether it is ever observed or calculated or not.

We cannot be concerned with the view that the universe itself is a closed system and therefore doomed beneath the advance of chaos' dread empire to unending darkness. Although only the entire universe *could* be a perfectly closed system, one doubts whether the statement 'The universe is a closed system' even makes sense, and there remains the cosmological enigma of its expansion continually enlarging the cosmic sink into which the material energy of its radiation unceasingly pours.[55] We can peer into these vistas of conjecture as long as we like. All that can really be said is that particular regions of the universe are sufficiently isolated for their inflow of heat to be so negligible that the Second Law with all its consequences must apply to them, and that a general increase of entropy necessarily results from present material processes. The laws of nature are not sacrosanct; they simply generalize the way the world behaves. To

[53] Op. cit., p. 95.
[54] which for Eddington, loc. cit., is a paragon of subjectivity.
[55] H. Bondi, *Cosmology*.

suppose them either extrapolated to infinity or abrogated is idly to imagine another sort of world altogether.

Yet neither universal increase of entropy nor universal expansion gets to the bottom of the matter. Even if neither existed, even if both were reversed and Eddington's 'arrow' with them, negentropy inaugurating a new world and the universe beginning to contract now as we write these lines, the past would be as extinct as ever, and before and after just as irreversibly before and after as they always have been. That Eddington did not see this is clear enough from his talk of 'events' as entities encountered on our way through time;[56] for if (with Mr J. W. Dunne as our unavoidable travelling companion) we meet them in a certain order going one way, we could expect, were the strange destiny that pushes our consciousness inexorably onwards in time somehow revoked, to find them there *in reverse order* on our way back.

This hypostatization of events is not peculiar to Eddington, but is in one form or another characteristic of much would-be scientific philosophy. Russell, for example, although he allowed them only brief duration, called events 'the *stuff* of the world',[57] and there are probably still those who would argue that such an event as 'my death' changes because, moving nearer in the future, it will one day become present and thereafter recede into the past. 'And from Everlasting to Everlasting those Things were in their Times and Places before God. . . . Like men in a ship we pass forward, the shores and Marks seeming to go backward, tho we move, and they stand still.'[58]

These picturesque ways of thinking lead directly back to time spatialized as a fourth dimension. As Prof. Smart has pointed out,[59] events are not things but happenings to things, their acting or moving or changing or becoming other things; events cannot themselves act or move or change or become anything. Strictly speaking, they do not even happen,[60] although Smart allows them to do so; they just *are* happenings, and happenings themselves cannot happen any more than motions can move. *But they do have a temporal order, and this order is absolutely irreversible.* We are not referring to observers and their *observations* of events, although these too have irreversible temporal order, but to events themselves. It is by juggling between ob-

[56] E.g., *Space, Time and Gravitation*, p. 51, quoted by Prof. J. J. C. Smart in his paper 'The River of Time', *Essays in Conceptual Analysis*, ed. Flew.

[57] *History of Western Philosophy*, I, ix; his italics.

[58] Traherne, *The Fifth Century*, 8.

[59] Loc. cit.

[60] Dunne makes this point in *An Experiment with Time*, VI.

servations and events themselves that the familiar relativity-tricks with temporal order are performed. Observations are necessarily relative to their observers, events are not.

That events have an irreversible temporal order implies that although happenings may be indefinitely repeatable, the clock striking noon today as it did yesterday, their repetitions are always *different events*. Even eternal recurrence would be the interminable rotation of a measureless clock, each recurrence a world-age later.

Things change, and all their changes are changes of state. A state is a qualitative, quantitative, formal and positional totality. A thing changes *from* one state *to* another state. Its changing is an event. The thing may change again to the same state as before, but the change is always a new event; every change of state is a new event. Different kinds of change may together produce a new state, but the resulting change of state may be conceived as a single event. Changes of state therefore constitute a progressive series. The relation between those changes is therefore asymmetrical and transitive; that is to say, their order is irreversible. By the same figure of speech that applies to other progressions, for example the series of natural numbers, the relation between them is called, by spatial metaphor, 'before or after'. Visualist misconception of the progression of change is much reinforced by the actual spatiality of positional change. The progression of motion *is* spatial.

Time, the 'phantasm of before and after in motion' as Hobbes called it, is the 'before and after' progression of change quantified by regarding any one process of motion or change as standard and rating other changes or relative stabilities against it, reckoning them in its terms. Unconscious awareness of our own cellular processes gives us our 'time-feeling', reference to external processes gives us our objective times; and realization of the standard process as a series of events that can be progressively numbered gives those times their useful measurement. 'For time is just this,' said Aristotle: 'the counting of motion in respect of before and after.'

Which process is to be regarded as standard is, of course, purely a question of convenience. The obvious recommendations are: observability, which enables the process to be referred to; continuity, which enables it to be artificially divided; repetition, which makes it divide itself; and an apparent regularity which, if accepted as perfect, determines the irregularity of other processes. Hence all the clocks mankind has ever used, and their eventual divergence.

There are just two corollaries, one apiece for the past and the future.

For the first, Comrade O'Brien arises again to confront us with the slogan of 1984: 'Who controls the present controls the past.'[61] If, in fact, the past no longer exists, if the earth upon which that past occurred has, relative to the rest of the universe, moved spatially beyond the very places of its occurrence and left a void behind, what except mere technical impotence is to prevent totalitarianism from so suppressing and falsifying its evidence that no one will ever again be able to discover what it really was?

The only answer to this is that the world of 1984 is a negative utopia with the incredible fixity of all utopias, and that the inexhaustible dialectic of history itself forbids Leviathan his final synthesis. Moreover, the nightmare is beside the point. There is, surely, no need to waste a couple of pages on the half-baked Bergsonian idea,[62] which reappeared in Alexander's *Space, Time, and Deity*,[63] that the past exists in memory. The visualist fallacy has been exposed often enough.[64] Occultists excepted, even those who believe that the past is still somewhere do not pretend to return there to check the truth either of present records of it or of their memories: like the rest of us, they can only compare them with other memories, other records, and all alike are fallible. What really happened may not be discoverable, its records defaced or vanished and its memory lost. So much the worse for our understanding of the present. False histories misshape our intellect, but they cannot misshape the past, for it exists no longer, and no violence either of earth or of heaven can alter its deceased reality.

The second question is that of 'precognition'. The past really existed, it developed into the present; by the paradox of light's limited velocity, if we are far enough away from where it occurred we can even now see it happening, for all that it exists no longer. But the future does not exist yet, the very spaces where it is going to happen are still to be reached, they are a void yet to be filled. Except by inference from what happened in the past or is happening in the present, the future is absolutely unknowable.

How was it, then, that in London during the Second World War a professional photographer named Basil Shackleton scored so successfully in a long series of card-prediction experiments that he became known as 'the man two-and-a-half seconds ahead of time'?[65]

This cannot be so shortly answered. We are about to cross strange frontiers.

[61] George Orwell, *Nineteen Eighty-Four*, III, 2.

[62] *Matter and Memory*, III.

[63] I, iv.

[64] E.g., in Russell's *History of Western Philosophy*, III, xxviii.

[65] S. G. Soal and F. Bateman, *Modern Experiments in Telepathy*.

4

First, the evidence must be marshalled.

Before Rhine it was mere narrative, ranging from old wives' tales to such tours-de-force as, to take what must be a very fair pick, Dunne's Martinique and 'Flying Scotsman' dreams,[66] the three cases chosen by Whately Carington (from H. F. Saltmarsh's collection of three hundred and forty-nine) for citation in his article on 'Precognition' in the 1955 edition of *Chambers's Encyclopaedia,*[67] and Dr S. G. Soal's Gordon Davis case.[68] Then in 1933 Dr J. B. Rhine in his newly-founded Parapsychological Laboratory at Duke University for the first time in history made it objectively statistical. His success inspired others, notable results being obtained by G. N. M. Tyrrell in 1935,[69] by Whately Carington in 1939,[70] and finally by Dr Soal's work with Shackleton in 1941 and 1942.

This evidence, familiar enough to amateurs of the subject—for the *good* evidence, far scarcer than generally supposed, is frequently recounted—may be summarized as follows.

(1) In 1902 J. W. Dunne, on camp in the Orange Free State, dreamed that a volcanic island was about to explode. He had dreamed of this island before, but more vaguely. This time he was vividly aware that it was French, with four thousand inhabitants. He awoke from a nightmare of trying to arouse the local authorities to the danger. The next post brought his *Daily Telegraph* with the first news of the Martinique disaster, stated to have caused over forty thousand victims. Still under the impression of his dream, Dunne read '40,000' as '4,000', and did not discover his mistake until he looked up the original report fifteen years later. Actually, as the next issue of the newspaper made clear, the number of victims was quite different from either figure.

(2) The 'Flying Scotsman' dream in the autumn of 1913 was of a place that Dunne recognized as being near the Firth of Forth Bridge, a high railway embankment with grassland below in which little groups of people were moving about. Suddenly he saw that a train had fallen down the embankment. Used to the idea that his dreams were often prophetic, he tried to get the date, and thought it was in

[66] *An Experiment with Time*; the 'Flying Scotsman' dream also in Soal & Bateman, op. cit.

[67] The second one is retold in Dr D. J. West's *Psychical Research Today.*

[68] Soal & Bateman, op. cit.; also West, op. cit.

[69] Idem.

[70] Idem.

the following spring; his sister, whom he told about it, subsequently remembered that he mentioned March. On April 14, 1914 the 'Flying Scotsman' went off the rails fifteen miles north of the Firth of Forth Bridge and fell to the golf-links twenty feet below.

(3) A lady saw an apparition of her mother-in-law. Some hours later she had an auditory hallucination: 'Come both of you on the Twenty-Second!' She went. The mother-in-law died on the 22nd. At the time of the hallucinations the lady had not known that she was ill.

(4) On July 18, 1882 a Mrs Schweizer dreamed that she and her son and a stranger were walking on a cliff and that her son slipped over the edge; the stranger said his name was Henry Irvin. On July 26th the son, riding with a friend on Scarborough cliffs, fell off his horse and died. Mrs Schweizer met the friend and recognized him as the stranger of her dream. His name was Deverell, but when she recounted her dream he told her that in amateur theatricals he had in fact used the pseudonym 'Henry Irvin'.

(5) A lady dreamed that after reading family prayers she went into the dining-room and found a pig there. She told her family about it. Upon going into the dining-room after prayers, there was the pig. It had been safe in its sty at the time of her dream.

(6) In 1922 Dr Soal was at a seance when a spirit announced itself as 'Gordon Davis' and gave sufficient personal details to be recognized as an old schoolmate whom Soal had last met by chance during the war and whom in 1920 he had heard of again as killed. At another seance five days later the medium's 'control' gave a description, which was noted, of the house in which Davis had lived.

That was all until 1925, when Dr Soal discovered that Gordon Davis was happily alive. He was living in Southend in a house that corresponded most remarkably with the description given by the medium three years before. But Davis had not been living there at all at that time, although between the first and second seances he had gone to look over the house, his furniture being in storage in London.

The suggestion is that the medium 'precognized' the following circumstances: that the street in which Davis was to live was not a proper street, but like half a one (it faced the sea); that its name began with two E's (Eastern Esplanade);[71] that the house was to have a 'dark tunnel' (a passage to the back); that there was to be a 'verandah' opposite (a promenade shelter);[72] that the pictures were to be of

[71] According to West, op. cit., what the medium actually said was: 'Get the letter E.'

[72] Oddly enough, West's apparently verbatim account of the medium's utterances reads: 'There's something right in front of the house—*not* a verandah.' Cf. Soal & Bateman, op. cit., p. 76.

mountains and the sea, and one of a road between two hills;[73] that there was to be a stuffed bird on the piano, two 'saucers' on the walls,[74] and two brass candlesticks downstairs. Dr Soal somehow calculated the odds against all these correspondences being due to chance and found them to be 'several million to one'.

(7) Dr Rhine's precognition experiments began when he asked his star subject Hubert Pearce to forecast the symbol-order of a pack of twenty-five Zener cards. After writing down his list of guesses, Pearce shuffled the cards, and the resulting order of their symbols was then checked against his list. Since the pack contained five sets of five different symbols, chance guessing should have averaged 5 scores a pack. Instead, Pearce's average score over 16 runs was 7.7 a pack. The odds against this being due to chance were about 10,000,000 to 1. Rhine then organized a series of 113,075 trials with 49 subjects guessing the future order of the cards and 11 experimenters doing the shuffling. The odds against the results being due to chance were over 100,000 to 1.

The subsequent development of precognition experiments at Duke University will appear later.

(8) G. N. M. Tyrrell built a machine consisting of a row of five lidded boxes each containing an electric lamp connected to one of five keys on the other side of a screen. He put a commutator in the circuit so that he could not know which key corresponded to which lamp, and a delay-action mechanism so that the actually corresponding lamp would not light until the lid of its box was raised. Mr Tyrrell sitting behind the screen pressed a key, and his subject Miss Johnson on the other side of it had to guess the right box to open, her successes being mechanically recorded. Odds of 1,000,000,000 to 1 against the results being due to chance indicated Miss Johnson's clairvoyance of electric circuits.

The precognition tests were made by Miss Johnson opening a box half a second *before* Tyrrell pressed a key. The results of 2,255 trials had odds against chance of about 270,000 to 1.

(9) Whately Carington's experiments had 741 participants in Great Britain, Holland and the USA; the 'transmitting centre' was his study in Cambridge. Each day for fifty days he picked a number at random from a mathematical table; this number determined the page at which a dictionary was to be opened; from this page he selected the first drawable object, which was then drawn by his wife. Her drawing was pinned up in the study at 7 p.m., and left there until

[73] West's account has only: 'Lots of pictures—all scenes—glorious mountains and the sea'.

[74] On the walls? Cf. West: 'Some vases—very big ones with funny saucers.'

9.30 a.m. the next morning, during which time the guessers made their own drawings, without dating them. These drawings, posted at regular intervals to Cambridge, were given code numbers for their subsequent identification. At the end of the fifty days they were all thoroughly mixed and passed to an independent judge whose task it was to match them with the fifty originals. A method was devised for scoring and evaluating the results, and the odds against their being due to chance were finally estimated at about 1,000 to 1.

But when the identification numbers were decoded, it appeared that in a remarkable number of instances *the drawings resembled originals that had not yet existed.* That is to say, those who presumably had been doing their best to reproduce the drawing actually on show, had on several occasions unwittingly succeeded in doing something much more startling: they had reproduced a drawing that was not yet drawn, of an object to be decided at random a day or two later. They had also, incidentally, made drawings resembling originals actually shown an evening or two earlier. Telepathy, it seemed, was somehow 'not focussed very sharply with regard to time'.

(10) The story of how Dr Soal, after five years of fruitless attempts to repeat the statistical feats proclaimed by Dr Rhine in 1934, was by 1939 so discouraged as to conclude that they were an unexportable product of Duke University, of how Whately Carington then suggested that he review all his results for the temporal displacements that Carington himself had just found in the drawing experiments, and of how with admirable pertinacity he did so and was rewarded, has been often retold. Only two of his one hundred and sixty subjects, Basil Shackleton and Mrs Gloria Stewart, had delivered guess-sheets which even the revised basis showed as positive, but it was enough. The results of his renewed work with them are generally agreed to be the best statistical evidence for 'extra-sensory perception' (ESP) ever produced.

The long series of experiments with Mrs Stewart, highly successful though it was for the demonstration of straight telepathy, is here irrelevant, for she never made 'precognitive' scores again. For Soal this was 'a bitter disappointment',[75] as though Shackleton had not already made 'precognition', whatever it might turn out to be in reality, an issue of scientific importance.

The experiments with Shackleton were of two types, Prepared Random Numbers (PRN) tests and Counter tests. In PRN tests, the order in which cards were to be guessed was mathematically random, being determined by lists of random numbers (1 to 5) previously compiled from standard tables. Five different picture-cards (elephant,

[75] *Modern Experiments in Telepathy*, p. 209.

giraffe, pelican, lion and zebra) were laid out face downwards before an agent, their order corresponding to the numbers 1 to 5. The experimenter showed the agent a printed card bearing the random number next on the list, the agent lifted the corresponding picture-card and looked at it, and Shackleton sitting in the next room then wrote his guess. The average interval between calls was 2.8 seconds; the conditions were those of a scientific experiment.

The results of 3,789 PRN calls were astounding. Direct hits were at chance level, but correct guesses of the *next* card to be turned up were so abnormally frequent that their statistical probability was 1 in 10^{35}. Subsequent tests showed that when the agent did not *look* at the cards, Shackleton could no longer guess them; that is to say, precognitive telepathy appeared to be operating and not precognitive clairvoyance. If the speed of calling was doubled, his predictions shifted to the *second* next card, 529 calls of this kind resulting in odds against chance of about 1,000,000 to 1.

In the Counters tests, the experimenter had a bag or bowl containing two hundred counters of five different colours, and took them out one by one without looking; the colour of the one taken determined which of the five cards the agent was to look at while Shackleton made his guess. The odds against the predictions scored by 1,578 guesses being due to chance were nearly 10^{11} to 1. When 794 calls were made at the rapid rate. Shackleton's predictions again shifted to the second next card, with odds against their being due to chance of over 10^{8} to 1.

All this evidence is discordantly judged. We may disregard for the moment the growls of the orthodox scientist at the statistical case for ESP as a whole: 'Bloody bad science! . . . It's no good doing a lot of experiments first and then discovering a lot of correlations afterwards, not unless the correlations can be used for making new predictions. Otherwise it's like betting on a race after it's been run.'[76] Unfortunately, it does not appear that the orthodox scientist has ever cared to give the evidence scientific attention, and there are none so deaf as those that dare not listen.

The verdicts of parapsychologists themselves range from Carington's enthusiastic 'There can be no doubt that the evidence for precognition is far stronger than that ordinarily needed for the establishment of any scientific phenomenon, however novel',[77] to Rhine's less

[76] Fred Hoyle, *The Black Cloud*, pp. 178-179. Cf. H. R. Post, 'Retrospective Miracles or Betting After the Race' (*The British Journal for the Philosophy of Science*, February 1960), being a review of Rhine & Pratt's *Parapsychology*.

[77] *Chambers's Encyclopaedia*, loc. cit.

committal 'strongly suggestive',[78] for whereas Carington was fascinated by precognition, Rhine's pet phenomenon is psychokinesis or PK; and, as we shall see, much of what is generally believed to be evidence for precognition conceals PK instead.

In this strange new world of spectral ambiguity, the one safe rule is that nothing is what it seems to be. Take Carington's sentence again, say 'survival after death' instead of 'precognition', and you will find it equally persuasive; but the delusive spirit of Gordon Davis will gibber at you from Summerland.

5

On the assumption that precognition is real, the wildest hypotheses have been proposed to explain it. Nearly all of them rely on the spatialization of time in a fourth dimension, so can be ruled out of court at once.[79] Of the very few that remain after this mass exit, the only ones at all noticeable are those of Whately Carington, Dr G. D. Wassermann and Dr Robert Amadou.

Carington thought that since subjective idealism teaches that what we call material objects are really no more than bundles of sense-data, these sense-data may be supposed to be floating about separately as self-existent entities *before* they combine to produce material objects, so that it is theoretically possible to be aware of them at any time.[80] The kindest thing to do about this artless theory that the ultimate constituents of the world are unsensed sense-data is to regard it as the philosophical equivalent of a hiccup; the less said about it the better.

Dr Wassermann's original hypothesis[81] sounded, as befitted a mathematical physicist, more erudite, but still seemed to arise from some kind of metaphysical indigestion. It presumed an infinity of timeless mental patterns 'striving' for physical embodiment, rather like Whitehead's Eternal Objects in their Realm of Possibility achieving Concreteness. Awareness of these striving eternal patterns would give foreknowledge of their materialization. Little need be said about this,

78 *New World of the Mind*, p. 82. Cf. West's 'strongly suggested' in *Psychical Research Today*, p. 131.

79 For a summary of them, v. Soal & Bateman, op. cit., pp. 171-175. Soal's own conclusion is that they are 'confused thinking' and give rise to 'many muddles'.

80 *Chambers's Encyclopaedia*, loc. cit. Cf. the theory of C. C. L. Gregory and Anita Kohsen noticed and dismissed by Dr C. T. K. Chari in 'Quantum Physics and Parapsychology', *The Journal of Parapsychology*, September 1956.

81 Reported with approval by Soal & Bateman, op. cit., pp. 169-170.

either. The timelessly striving patterns of Dr Soal's picture-cards of giraffes and pelicans speak for themselves.[82]

Later, Wassermann repressed this shameless mentalism beneath an elaborate 'physicalistic' theory in terms of newly postulated fields, morphogenetic fields and behaviour fields and psi-fields, which he presented as a scientific correlation of ESP phenomena in general. He stretched it to include precognition by a corollary splitting those fields into duplicate pairs of which one made its transitions more rapidly than the other so as to give an earlier field-representation of future events.[83]

The perplexity with which the theory was greeted by his fellow-scientists indicated, although it is doubtful how far this was realized, a fundamental difference of principle as to what a scientific theory is intended to do. Stretching away beneath it was that hidden schism between the materialist and positivist attitudes which is undermining the integrity of Western science. The explicit reference to Carnap's physicalism should, of course, have aroused their suspicion at once, for that neo-classic façade of pseudo-scientific agnosticism can conceal the smokiest altars. In fact, all Wassermann claimed for his theory was that it incorporated ESP into the language-system of 'unified science'.[84] That is to say, he was not claiming to tell them something new about the universe, something they might discover to be really there; he was only suggesting how to talk about ESP in a new way, constraining the timeless mental patterns to strut in physical disguise. To wonder if the hypothetical fields actually existed, to ask how the theory could be established as the objective *truth* about ESP, was naively to mistake its purpose.[85] These are *metaphysical* questions;[86] and in the physicalist vocabulary 'metaphysical' means 'nonsensical'.

Although any demonstration that ESP can be interpreted in purely physical terms might, after all Rhine's missionary preaching of a new world of the spirit, be regarded as already a victory, a physical theory must be more than a convenient mode of expression. Science is not a convenient language; it is actual or potential knowledge. Either a physical theory conceives processes which it apprehends as physically existent and prescribes courses of action, however technically difficult,

[82] And, to paraphrase the defunct Reichsmarshal, when I hear the word 'striving' I reach for my gun.

[83] Wassermann, 'An Outline of a Field Theory of Organismic Form and Behaviour', *The Ciba Foundation Symposium on Extrasensory Perception*, ed. Wolstenholme & Millar.

[84] Ibid., pp. 201-202.

[85] Cf. ibid., pp. 92-93 and 129-130.

[86] Cf. ibid., p. 67n.

that would test its correspondence with physical reality, or it is speculative fiction, a useless castle of words.

Anyhow, Wassermann's theory gives no help to precognition. The faint plausibility of its corollary collapses when we realize that there is hardly a single practical instance of apparent precognition to which it could apply.

Dr Amadou's hypothesis[87] is derived from Jung, who owed it and much else to Schopenhauer.[88] It concerns personal presentiments only, and does not attempt so much to explain them as to explain them away, the idea being that what happens to us, however apparently accidental, is unknowingly stage-managed by ourselves. Our misfortunes are self-punishments, our strokes of good luck the rewards of secret conceit, we are the unconscious masters of our fates; and now and again we obscurely know what the backstage dramatist who is our own deepest self has prescribed. This shadowy notion, recalling the classical Brooks case[89] in which a death-premonition brought about its own exact fulfilment, may have something in it, but its restricted application makes it of no use to explain away precognition generally.

The truth is that any attempt to explain precognition as real fails before it starts. Those who believe that its possibility is indicated by the discovery that certain subatomic phenomena show a 'reversal of time', a bombarded nucleus appearing to explode an infinitesimal fraction of a second before it is hit, ignore the fact that, like other startling features of quantum theory,[90] this oddity arises from the inevitable distortions of microphysical observation; that is to say, it is an optical illusion due to the motion of subatomic particles at a higher speed than light.[91]

Precognition as knowledge of the future uninferable from present knowledge is an impossibility; and if it implies that the not-yet-existent somehow exists, it is an absurdity. There can be no explanation of an absurdity, and it is useless for parapsychologists to strain themselves to provide one.

An absurdity cannot be explained; it can only be explained away. That which is impossible must be resolvable into that which, how-

87 Ibid., 'Parapsychology in the Modern Approach to Psychosomatic Man', esp. pp. 196-197.

88 Somewhere in *Parerga and Paralipomena*, vol. I.

89 E.R.E., art. 'Presentiments'.

90 E.g., 'quantum-jumps'.

91 Cf. Prof. Hideki Yukawa's paper at the 2nd United Nations Conference on the peaceful uses of atomic energy, Geneva, September 1958. On the whole question of the interpretation of quantum theory, v. the Colston Research Society's Symposium on *Observation and Interpretation*.

ever unlikely, is at least possible. It is only in Aristotle's *Poetics*[92] that a plausible impossibility is preferable to an implausible possibility. Parapsychology, seeking scientific status, cannot afford such extravagance. Its persistent inability to provide a scientific explanation of the phenomena it investigates is, one hopes, a remediable shortcoming; but infatuation with an impossibility can only perpetuate its scientific immaturity. Inexplicability is merely contingent; impossibility is incorrigible.

All instances of apparent precognition must be reducible to hidden factors operating at the time of the apparently precognitive act. The process will often be tortuous, for the psi-world is as full of tricks as the original monkey, and even in the ordinary world true explanations seem more complicated than misapprehensions, natural causes harder to be understood than supernatural ones. This is the paradox of Ockham's Razor: the fewer your hypotheses, the longer your answers.

It would, moreover, be stupid to be surprised at improbabilities. These are, surely, what we expected to find in the psi-world, its proper fauna. None of them will be more unlikely than the complicated ESP theory that is the only real alternative to the discarnate spirit-possession hypothesis for explanation of the phenomena of mediumship; and even spiritualism, however incredible, cannot be branded with the impossibility that outlaws precognition. In short, we must be prepared for almost anything that is not downright impossible. To repeat the Aristotelian echo of *Finnegans Wake*, 'we are in for a sequentiality of improbable possibles.'[93]

These have such labels as clairvoyance, telepathy, psychokinesis. We should prefer more comprehensible ones, but at present this cannot be helped. If later these obscure improbabilities can themselves be explained away, so much the better; for the moment it is important to eliminate only what is plainly impossible. If everything else fails, there is always chance—chance that makes your telephone number coincide with your car number, that made Mike Todd's aeroplane crash on Mount Taylor,[94] that made Some Baby romp home at 5 to 1 in the Tally-Ho Hunters' Chase at Lingfield on the day Queen Elizabeth II gave birth to Prince Andrew,[95] that made the cabalistic numbers of the total eclipse of the sun on February 15, 1961 come

[92] Ch. 24.

[93] p. 110.

[94] And the aeroplane was named 'Liz'. Shameless presumption that this book may outlast Elizabeth Taylor's film fame leads me to add that Liz was Todd's wife at the time.

[95] *New York Herald Tribune*, Paris, 22nd February 1960.

out in the next Saturday's extraction of the weekly Italian State Lottery.[96] [97]

As good an example of 'precognition' as most of the dreams related by Dunne in *An Experiment with Time* is one of my own in September 1957. I dreamed that I was introduced to a Mr Samuel Bowman and thought what an odd surname he had. The next day, discovering in the British Council Library in Rome a copy of Hugh Kenner's *Dublin's Joyce*,[98] I was struck by a reference to Isa Bowman, the precocious American child-actress of the stage version of *Alice in Wonderland* who supplanted Alice Liddell in the Rev. Charles Lutwidge Dodgson's affections. I had never heard of the creature before, or seen the book; and I wondered what odds against chance Dunne would have awarded the dream.

Unfortunately for precognition, I have here been emulating, although to the opposite purpose, the impishness of Dunne himself, when to increase the impact of an unlikely coincidence between a dream and a waking event he sometimes misleads his readers to believe that the dream occurred after the event before revealing that it was in fact the event that followed the dream. What really happened was that I had borrowed Kenner's book from the library, and browsing through it had found the reference to Isa Bowman. I had indeed never heard of her before, and thinking what an odd-sounding name it was, I imagined that a man of Dodgson's sensibility must have found it something of an obstacle; anyhow, I annotated my copy of *Finnegans Wake* accordingly. The next morning, U.S. citizen Samuel Bowman,[99] just arrived in Rome on an international coach tour, telephoned me from his hotel. He had found my address in his American insurance policy as the person to whom claims arising in Italy should be reported, and therefore wanted to let me know that he had injured his knee-cap in a sudden jolt of the coach. (I often get such calls from people unknown to me. Far too many have interrupted the writing of this book.)

The fecundity of coincidence! Examples could be multiplied from the experience of us all. However, not much need be left to chance; ESP and PK between them will be generally sufficient. Let us, then, return to the best evidence for precognition and see how it fares.

[96] Cagliari wheel: 27 (eclipse), 8 (its hour), 37 (its minute). Much money was won on this popular 'terno' in Florence, where the eclipse was total; also 61 (the year) came out.

[97] and that involves two Elizabeths in the only two personal coincidences here cited.

[98] Chatto & Windus, 1955.

[99] I will not cheat and say his name was Isaac.

6

(1) *Martinique Dream*

Dunne was evidently a man of lively curiosity, a devourer of the daily news, and it seems to have made his dreams more variedly impersonal than those of most people. The fate of Krakatao had impressed him, and he had already had several dreams about a volcanic island in danger. He himself admits the possiblity of telepathy with the *Daily Telegraph's* journalists, although it might have been with anybody, after the worldwide stir of the Martinique disaster. But the more likely explanation would be clairvoyance, not of Martinique, of course, but of the newspaper he was waiting for, with such avidity that he apparently succeeded in reading its headlines before it reached him. This is a highly improbable feat, but nothng like so incredible as his suggestion that the dream was caused by his *future* reading of those headlines.[100]

(2) *'Flying Scotsman' Dream*

This is the best of the lot, perhaps the one occasion where we might have to appeal to the fecundity of coincidence. We can hardly suggest that Dunne sent the train off the rails by psychokinesis, and to suppose his clairvoyance of a defect in the line appears desperately far-fetched, despite the old case of the negro preacher who, dozing alone in his compartment on a Washington-bound train, was awakened by a repeated cry of 'Wreck! Wreck!', whereupon he rushed to the end of the coach and jumped off: nothing happened to *that* train, but the next one going through was wrecked by the falling of a large rock that overhung the line.[101]

If this story (the negro one) is true at all,[102] it is surely explicable by clairvoyance of the dangerously loosened rock producing an hallucinatory warning, but here the negro was personally involved. On the other hand, Dunne was familiar enough with that railway embankment. We wonder if he had dreamed about railway accidents before, or if there had been a recent report of one in his newspaper. There are, indeed, many things we do not know that might make the coincidence less extraordinary than appears from what he published.

The great objection against premonitions in general is their poor evidential status. Innumerable unknown ones remain without conse-

[100] Cf. *An Experiment with Time*, 6.

[101] T. J. Hudson, *The Law of Psychic Phenomena*, XVII.

[102] It was related to Hudson by one who had seen the old preacher arrive cut and bruised at the next station after walking along the line.

quence, and asserted fulfilments lack proper corroboration. As long ago as 1888, after the unsatisfactory Schweizer case, the Society for Psychical Research published an appeal that people having dreams or impressions which they thought might be premonitory report them to the Society *immediately*, so that their eventual fulfilment could be properly checked. When the Society's Experimental Research Officer went through the files sixty years later, he found only thirty-two such premonitions on record. The events they forecasted were all quite possible by chance alone, but none of them had come true.[103] Against Dunne's dreams in particular stand the unfortunate facts that he did not take this or any other opportunity to establish his claims, and that serious attempts to repeat his results have been unsuccessful.[104]

(3) *Mother-in-law Case*

She was already ill at the time of the apparition. A clear case of telepathy. The death on the 22nd suggests Amadou's theory, but might well have been unconsciously inferred.

(4) *Schweizer Case*

It is clear from the fuller account given by West[105] that Mrs Schweizer did not *immediately* recognize Deverell as the stranger of her dream. The resemblance *dawned upon her* while she was sitting opposite him in the train during their journey together to visit the scene of her son's death, and it was *then* that she told him about her dream and was told by him that he had sometimes used her dream-stranger's name. No proof that he had in fact done so was ever obtained.

Although Amadou's theory could provide a complicated explanation, the evidence is so defective as to make it unnecessary. Mrs Schweizer, an ardent spiritualist, did not report the case to the Society for Psychical Research until three months afterwards. Her statement was countersigned by her other son John. She declared that a similar statement countersigned by Deverell had been sent to a spiritualist magazine, but the document was not traced, and Deverell was drowned before the investigators could obtain any communication from him.

As Dr West remarks, the more striking the case, the worse the supporting evidence. The story rests upon the belated testimony of

[103] West, *Psychical Research Today*, pp. 38-39.

[104] Besterman, 'Report of an Inquiry into Precognitive Dreams', *Proceedings of the S.P.R.*, 1933, referred to by West, loc. cit.

[105] Op. cit., pp. 37-38.

Mrs Schweizer who was subject to continual hallucinations[106] and of her son John who was a practising medium.

(5) *Pig Case*

This is a silly story. The household was a large one (Carington says a palace) and the dream had been talked about beforehand. The pig was simply taken out of its sty and put into the dining-room by an undiscovered practical joker to make the dream come true.

(6) *Gordon Davis*

This is one of the most interesting cases of ESP ever recorded, but it is hard to see why it is supposed to involve precognition. The medium did not, after all, describe the house until Davis had been to look over it with the intention of living there, and it was natural enough that he should at the same time have pictured the arrangement of his stored furniture.[107] The coincidences regarding the setting of the ornaments are anyhow unimpressive; what is important is the description of the house, and this could be precognitive only if Davis had never been there before. The medium seemed to Soal to be foreseeing it all because he himself did not discover Davis and see the house until three years later. But the medium had inscrutably discovered them before she spoke; and since Soal believed Davis to be dead, she characteristically produced him as a spirit communicating from beyond the grave.

To get the full flavour of this Gordon Davis case, it should be read beside the equally curious John Ferguson one,[108] too long to be recounted here, in which during a long series of seances with the same medium Mrs Blanche Cooper, she and Dr Soal unconsciously composed between them the life and death of a man who, it finally turned out, had never existed, but whose air-built biography projected astonishing highlights of extrasensory perception. As the spirit of Gordon Davis remarked when introducing itself: 'The dead to the living! Queer world, what?'[109]

(7) *Dr Rhine's Experiments*

Rhine himself was quick to see that instead of his subjects having forecast the future order of the cards about to be shuffled, his experimenters might just as well be using ESP to shuffle them into the order prescribed by his subjects. This much less startling hypothesis was am-

[106] She had direct conversations with spirits and her dead husband accompanied her when she went out for walks.

[107] West, op. cit. p. 73, does not mention the furniture as already existing in storage, but Soal himself, op. cit. p. 76, says it was so.

[108] Soal & Bateman, op cit., pp. 260-264.

[109] *The Journal of Parapsychology*, December 1960, p. 267.

ply confirmed by a new series of experiments in which 203 participants made a total of 211,525 attempts to shuffle their Zener cards to match concealed test-packs, and succeeded to a similar statistical extent as in the supposed precognition tests. The odds against chance were over 180,000 to 1.[110]

Precognition was, therefore, back where it started. In the hope of setting it off again, Rhine decided to have the cards shuffled by a machine. The subject made his list of twenty-five guesses while an experimenter turned the handle of a mechanical shuffler containing fifty cards; when the list was ready the experimenter stopped, and the first twenty-five cards in the machine-shuffled pack were compared with it. The correct guesses in a total of 235,875 had odds against chance of only 30 to 1.

Another seven series of experiments totalling 154,675 guesses were then carried out, still using a machine but with a different procedure. The results have been much disputed. Rhine claimed victory with a total of 509 scores beyond the average producing odds against chance of about 800 to 1. But, as Dr Soal has pointed out, he obtained his total of 509 rather too ingeniously. In two out of the seven series of these experiments, 27,700 guesses had been made by adults and 12,500 by children. The children scored 123 *more* than the chance-average and the adults scored 239 *less* than the average. On the ground that preliminary tests had made him expect below-chance scores from the adults, Rhine was inspired to count their 239 *misses* as *hits*. If they are counted as misses, the odds against chance of the entire set of 154,675 results are reduced to zero.[111]

The only consolation parapsychological statisticians were able to draw from these experiments, which were completed in 1940, was that in the two crucial series the difference between the scores of the adults and the children was sufficiently improbable to indicate a non-chance factor. By 1941, however, Rhine's interest in psychokinesis led him to recognize that his results were anyhow useless as evidence for precognition. The non-chance factor, if any, could always have been a combination of ESP and PK, the former making his subjects aware of the card-order inside the machine, and the latter enabling them to influence the shuffling to the very small extent required to produce a statistical effect. That is to say, the prophets would unconsciously have arranged the fulfilment of their own prophecies.

It was obviously necessary to eliminate any possibility of PK. This was done by having the number of turns of the shuffling-machine and also a final cut of the cards determined by the local maximum and

110 Soal & Bateman, op. cit., pp. 77-78; Rhine, *New World of the Mind*, p. 80.
111 Soal & Bateman, op. cit., pp. 79-80.

minimum temperatures of the previous day. A total of 2,302 lists of 25 guesses were submitted by 51 subjects, making 57,550 forecasts in all. About half of them were tested two days later and the other half ten days later. In both cases the scores were at a purely chance level.

Undaunted by this manifest failure, Dr Rhine searched doggedly through the results for any statistical eccentricity at all. For example, did the lists generally have more correct guesses in their first and last sections than in the middle? No, they did not, but he finally discovered what he called a 'salience-ratio covariation' and declared that the odds against its being present by chance were about 500 to 1.[112] Dr Soal considers this 'very recondite relation' a mere statistical artifact. As he observes,[113] by making all sorts of arbitrary subdivisions it is possible to find some abnormality in practically any collection of data, if only one tries hard enough. When, purporting to squash Mr Spencer Brown's claim that statistically significant results could be obtained simply by matching sets of random numbers one against the other, the Society for Psychical Research published the results of 98,000 trials with random numbers taken from the standard tables and declared that except for a slight decline effect every test upon them gave only chance results, Brown promptly discovered that they showed a pronounced Quarter Distribution (QD) effect,[114] one of Rhine's favourites, with odds against chance of about 5,000 to 1.[115]

Attempts to obtain significant results with the temperature-method continued at Duke University until 1954 or so,[116] but after this it appears to have been tacitly abandoned in favour of a method known as Mangan's target-determination procedure which did, very occasionally, produce encouraging scores. *The Journal of Parapsychology* from December 1957 to March 1962 reported one small series of 91 trials with a single subject giving results that had odds against chance of over 1,500,000 to 1,[117] a series of 18,300 trials with a class of schoolchildren that resulted in odds against chance of about 150 to 1,[118] and

112 Rhine, op. cit., p. 18.

113 Op. cit., p. 82.

114 An uneven distribution of hits between the four quarters of a score-sheet.

115 G. Spencer Brown, *Probability and Scientific Inference*, Appendix II.

116 Rhine, op. cit., p. 82. From the sources available to me it does not appear that the results were ever published.

117 December 1957, G. L. Mangan, 'An ESP Experiment with Dual-Aspect Targets involving One Trial a Day'.

118 September 1959, Margaret Anderson and Elsie Gregory, 'A Two-Year Program of Tests for Clairvoyance and Precognition with a Class of Public School Pupils'.

a remarkable series of 1,125 trials with a single subject forecasting Zener symbols *a whole year before* they were determined, that resulted in odds against chance of 500 to 1, although a parallel series of trials with the same subject on forecasts tested after five days gave only chance scores, so that the total results of both series were reduced to the insignificant odds against chance of 70 to 1.[119] Other experiments, some of which were published (San Diego Precognition Test 31,000 trials,[120] Wheaton Precognition Test 8,950 trials),[121] were failures.

Mangan's procedure consists in obtaining four three-figure numbers by four throws of three ten-sided dice, multiplying those numbers together, multiplying the product by itself read backwards, and taking the square root of the result. The last six figures of the root determine a place in a standard table of random numbers, and the random number thus obtained determines the target.

Now this sounds terrific, and a calculating machine is of course used. Yet there are many instances of mathematical prodigies accomplishing such feats. Zerah Colburn, for example, a child aged seven to whom 'the attention of the philosophical world' was attracted in 1812, 'will tell the exact product arising from the multiplication of any number consisting of two, three, or four figures by any other number consisting of the like number of figures; or any number of six or seven places of figures being proposed, he will determine with equal expedition and ease all the factors of which it is composed. This singular faculty consequently extends not only to the raising of powers, but to the extraction of the square and cube roots of the number proposed, and likewise to the means of determining whether it is a prime number, for which case there does not exist at present any general rule among mathematicians. All these and a variety of other questions connected therewith are answered by this child with such promptness and accuracy (and in the midst of his juvenile pursuits) as to astonish every person who has visited him.'[122]

[119] June 1959, Margaret Anderson, 'A Precognition Experiment Comparing Time Intervals of a Few Days and One Year'.

[120] December 1958, Margaret Anderson & Rhea White, 'A Survey of Work on ESP and Teacher-Pupil Attitudes', pp. 254-255.

[121] Ibid., pp. 260-262.

[122] The English *Annual Register* of 1812, extensively quoted by Hudson in *The Law of Psychic Phenomena*, pp. 64-67. Hudson believed intuition of 'the fixed laws of nature', in which he included the laws of mathematics and harmony, to be an inherent function of the 'subjective mind'. West, in *Psychical Research Today*, p. 63, is certainly wrong when he ascribes the feats of mathematical prodigies to one-track memorizing: he should read the evidence.

Prodigies exhibit the latent capacities of all. If, therefore, the scanty successes of Mangan's procedure *had* to be explained as other than statistical artifacts, which the absurdness of forecasting the order of Zener symbols a year in advance indicates them to be, here would be a possible basis of explanation.[123] And as long as any possible explanation exists, the impossible explanation of precognition remains excluded. The inadequacy of Mangan's procedure was implicitly recognized by an appeal in *The Journal of Parapsychology* in September 1958[124] for the renewal of experiments with the temperature-method, but it does not appear to have borne any fruit.

In conclusion, Dr Rhine's own verdict upon precognition in 1957, after twenty-four years' work to establish it, was *not proven*: the evidence was 'highly indicative' but still needed the 'collection of further data'.[125] And the insufficient evidence to which he was referring included that of Basil Shackleton.

(8) *Tyrrell's Experiments*

Parapsychologists themselves discount these because Mr G. W. Fisk demonstrated that a simple guessing system produced similar results, which so disheartened Miss Johnson that her scores collapsed to chance level.[126] Tyrrell himself proposed the explanation that ESP might have enabled him to know which box she had predicted and which was its corresponding key, but since Tyrrell had never manifested ESP abilities it is perhaps more likely that Miss Johnson influenced him by telepathic suggestion to press the right key. Precognition fares badly anyhow.

(9) *Carington's Drawings*

Carington believed that he had discovered the parapsychological phoenix, a successfully repeatable experiment; but when the American Society for Psychical Research did carry out repetitions of it, the results were unsatisfactory.[127] However, we are not arguing about the telepathic or clairvoyant communication of drawings, for which there is other evidence besides Carington's;[128] what we are con-

[123] Rhine, discussing the hypothesis that PK results may really be precognitive (op. cit. pp. 82-85), allows the possibility of a far more prodigious mental arithmetic than would be required to deal with Mangan's procedure.

[124] Pp. 216-217, William H. Clark, 'A Practical Application of Precognition'.

[125] *The Journal of Parapsychology*, December 1957, p. 301, review of *Parapsychology: Frontier Science of the Mind* by J. B. Rhine & J. G. Pratt; v. also *New World of the Mind*, p. 78.

[126] West, op. cit., pp. 105-108.

[127] Soal & Bateman, op. cit., pp. 91-92; West, op. cit., p. 102.

[128] E.g., Upton Sinclair's; v. West, op. cit., p. 101 and the illustration facing p. 96.

cerned with is the preposterous assertion that such communication can occur *before the drawings exist.*

Although Carington claimed that his 'precognitive' scores were *independently* significant, this has been much disputed, and its disputability lays bare what Dr Soal called the Achilles' heel of all his results: the unreliability of the method he used to evaluate them.[129] But even if they are valid, they are far from compelling belief in precognition. To assume that some of his participants were able, after clairvoyance of his number-table and his dictionary, to influence his choice to correspond with the drawings they had already made, or alternatively that he was telepathically aware of what they had drawn and so was unconsciously moved to select appropriate targets, may well strain credibility; but to suppose that his participants were able to copy drawings that did not yet exist breaks it completely.

(10) *Basil Shackleton*

The infinite suggestibility of the 'subjective mind' is an old psychical research commonplace unaccountably forgotten since psychical research became parapsychology. Personality is performed by consciousness, and self-consciousness even makes us our own playwrights; but in our unconscious depths we are born actors all, non-persons with limitless capacities of impersonation.

The one safe rule in the psi-world is, as we have said, that its phenomena are not what they appear to be. We may now add that what they appear to be is, however mysterious to the conscious intelligence, always the result of a hidden interplay of suggestion between their producers and the beholders who themselves participate in their production.

When Rhine pointed out that instead of Shackleton being absurdly '2½ seconds ahead of time' all his PRN results could have been obtained by clairvoyance of the next number on the list and hence of the card that corresponded to it, Soal replied that Shackleton never succeeded in clairvoyance tests. This reply begs the whole question as to what it is that decides the particular forms of psychic manifestation. Clairvoyance and telepathy are not faculties that may or may not be separately possessed. They are differentiated forms of the ESP-response evoked by certain inter-personal situations, and the form that response takes is profoundly determined by the suggestional complex of the situation itself.

Although Dr Soal does not seem to know it, he has never really cared for clairvoyance, so much so that he earned Rhine's reproach for overlooking the possibilities that could have made his work with

[129] Soal & Bateman, op. cit., p. 94.

the Jones boys an indisputable success.[130] Clairvoyance and psychokinesis were the magic of his rival wizard across the Atlantic, and Soal had his own name to make. His experiments were inevitably *Modern Experiments in Telepathy*, his subjects *The Mind Readers*.

Similarly, he liked precognition more than straight telepathy, and was genuinely disappointed when Mrs Stewart, instead of repeating the 'precognitive' scores he had discovered in her earlier results, was spurred by unconscious jealousy[131] of Shackleton to make parapsychological history of her own by switching her ESP to direct hits. Although her straight telepathy scores were fully twice as brilliant as Shackleton's 'precognitive' ones, Shackleton was always his favourite.[132] In short, had Ouspensky not already done so, Dr Soal could have entitled his work *In Search of the Miraculous*.

As for the Counters tests that were intended to meet the clairvoyance objection, Soal himself recognized that they were inadequate.[133] To repeat his own criticism, which Rhine endorses,[134] since the agent used alternate hands to take the counters from the bowl, she had at the normal rate already selected the next one by the time Shackleton made his guess; and at the rapid rate, which anyhow produced results 1,000 times less successful than those of the normal rate and 10^{27} times less than those of the PRN tests, it was legitimate to assume that Shackleton was influencing the agent's choice of counters to correspond with his calls.[135]

The best evidence for precognition is just not good enough.

7

Philosophers notoriously betray themselves in the last chapter; the

[130] Soal & Bowden, *The Mind Readers*, Appendix O (Comment by J. B. Rhine), p. 287. The criticism is amplified in Dr J. G. Pratt's review of the book in *The Journal of Parapsychology*, March 1960.

[131] Soal's own explanation: *Modern Experiments in Telepathy*, pp. 322-323.

[132] The very frontispiece of *Modern Experiments in Telepathy* is Shackleton's portrait alone.

[133] Op. cit., pp. 162-165.

[134] *New World of the Mind*, p. 78.

[135] In any case the choice of counters, especially at the rapid rate, could hardly have been statistically random. West, op. cit., p. 110, makes this his main objection to the Counters tests; cf. Soal & Bateman, op. cit., pp. 154-155. Incidentally, Soal's remark (op. cit., p. 163) that the above explanation of the rapid-rate tests requires, when the counters were taken from a closed bag, the additional hypothesis of clairvoyance by the agent, is incorrect. It is clear from pp. 154 and 156 that in *all* rapid-rate tests the counters were taken from an open bowl.

shrine at the end of the garden is a familiar surprise. No such object is to be expected here, but it remains to be answered why a book written from an avowedly materialist standpoint admits psychic phenomena. The answer is that they are admitted because they cannot be excluded.

World-pictures that feign completeness by omitting incongruities may be materialistic, but they are certainly not scientific. A century ago Engels could fail to see the implications of the hypnosis effects observed during his own curious excursion into psychical research[136] and retain his philosophical integrity. Today, the existence of unknown causal factors incongruous with the present conceptual structure of scientific materialism is upheld by objective evidence. Whatever the reality of those factors, the evidence for their existence can no longer be scientifically ignored. Recalcitrant materialists may be relieved to know that, much to Dr Rhine's mortification after years of pleading for U.S. Government funds to reinforce parapsychology as an ideological weapon against communist materialism,[137] the Soviet Union is the first country in the world to have instituted a State-supported parapsychological research laboratory.[138]

The only serious objection to the objective evidence for psi-processes is that it is entirely statistical. Much of the incredibility of the results we have been discussing fades when we realize that the enormous odds against their being due to chance are the large accumulation of slight deviations from a statistical norm. No single score presents itself as definitely psi-caused; all that can be claimed is that the total scores are so improbable that unknown causes are their only possible explanation.

Against this the orthodox scientist declares that abnormal statistics cannot establish abnormal causes, and in the last resort he declares probability theory itself to be unreliable: the exceptional can always occur. True, insurance companies make profits, and citizens of Monte Carlo have paid no taxes; but reinsurance exists precisely because loss-ratios are occasionally overwhelmed, and the man who broke the bank at Monte Carlo is music-hall history. Meeting Mr Smith by pure chance may be most improbable, but its happening is not a Psmith phenomenon.[139]

Now this argument is good enough for a raid into the para-

136 'Natural Science in the Spirit World' (*Dialectics of Nature*).

137 'Why National Defense Overlooks Parapsychology', *The Journal of Parapsychology*, December 1957.

138 *Parapsychology Bulletin*, May 1961.

139 H. R. Post 'Retrospective Miracles or Betting After the Race', *The British Journal for the Philosophy of Science*, February 1960.

psychological camp, a skirmish and out again, but it cannot stand a pitched battle. It has made *The Journal of Parapsychology* raise its significance criterion from 50/1 to 100/1,[140] but however much we should like it not to do so we feel it break in our hands when we try to use it to knock the bottom out of parapsychology altogether. Some of the strange results produced by parapsychologists are obvious probability freaks—we have seen one or two of them in our discussion of precognition—but no scientist can seriously maintain that Soal's results with Shackleton and Mrs Stewart or those of the Pearce-Pratt experiment[141] are explicable as statistical artifacts, and indeed the last 'scientific' attack on the Pearce results has had to fall back on the disgraceful hypothesis of fraud.[142] Even Mr Spencer Brown, the free-lance paladin of the statistical artifact argument,[143] retreats under pressure to the safer position that 'if more of us could get results like Dr Soal there would be no difficulty at all in accepting a communication hypothesis', and that 'the only thing I have said is that I *am not sure that all of them* (parapsychological results generally), or even most of them, are evidence of telepathy and what not.'[144] This is fair enough, but if the argument cannot cover *all* cases its original promise to explain parapsychology away for us is not fulfilled. Repeated unexpected encounters with Mr Smith becomes increasingly sinister, and we demand their explanation.[145]

Basically, what is wrong with parapsychology is not its results but its inability to make them intelligible. Not only has it no control over the processes it investigates and no theory to explain them, but it cannot even give a proper definition of them. 'No definition, and especially no operational definition, of ESP can be given.'[146] All that can be said is that the statistical results are highly significant, which is interesting, but inadequate. Significance means practically nothing until we know what it signifies. 'If we ask the question "What is ESP?" all serious parapsychologists will be unanimous: no answer

[140] Cf. the definitions of 'Significance' in the Glossary for 1956 and 1957 respectively.

[141] Soal & Bateman, op. cit., pp. 30-31.

[142] C. E. M. Hansel, 'A Critical Analysis of the Pearce-Pratt Experiment', *The Journal of Parapsychology*, June 1961.

[143] *Probability and Scientific Inference*. Cf. Christopher Scott's critique of it in the *Journal of the Society for Psychical Research*, June 1958, to which I was kindly directed by Dr J. G. Pratt.

[144] *Ciba Foundation Symposium on Extrasensory Perception*, p. 100, my italics.

[145] Cf. Spencer Brown's discussion of Mr X, *Probability and Scientific Inference*, pp. 38ff and 43ff.

[146] Amadou, *Ciba Foundation Symposium on Extrasensory Perception*, p. 191.

can be given.'[147] Until an answer *can* be given, parapsychologists can only go on ringing experimental changes to record more and more incomprehensible results. It is this failure to propose a scientific theory to explain its results that makes parapsychology fall short of scientific status and leaves it in the measurement and classification stage characteristic of a natural history. As long as its practitioners regard it as spiritual history, it can be expected to remain there.

Admittedly, parapsychology owes its very existence to the religious motives of its founders. It began when Dr Rhine, giving up the idea of joining the ministry, determined instead to devote his life to making the supposedly spiritual phenomena of psychical research a scientific issue. This has been achieved; but it is precisely the religious bias inherent in its whole development, its insistence upon the 'non-physical' nature of the phenomena it is investigating, that prevents parapsychology becoming a science. It is indeed fitting that this endeavour to demonstrate immaterial agency found its home in Duke University, the last refuge of vitalistic biology. What is basically wrong with parapsychology is its metaphysics.[148]

No science can admit the non-physical, for the non-physical is its antithesis. Phenomena are given a religious explanation until they receive a scientific one, and a scientific explanation necessarily removes them from the spiritual domain. Scientific explanations *are* physical explanations; the entire history of knowledge represents the overthrowing of non-physical explanations by physical ones. No scientific theory can be constructed with spiritual postulates, for they are scientifically sterile. In fact, parapsychologists *have* no such theory; all they can do is to stress that parapsychological phenomena defy physical explanation and to assert that this must denote their non-physical origin. As so often before, we are offered the old logical fallacy of *argumentum ad ignoratio*, 'this explanation is better than ignorance', when actually it is no explanation at all. Indeed, it is worse than no explanation at all, for it lulls ignorance with illusions of intelligence, and its unusual presentation in scientific garb is fundamentally an attempt against the whole scientific conception of the world.

Parapsychology cannot hope to become a science until it starts thinking scientifically, and scientific thinking does not consist in the

[147] Idem.

[148] This applies also to parapsychologists who are without religious bias. West, for example, in *Eleven Lourdes Miracles* doubts faith-healing, but in *Psychical Research Today* he accepts precognition, which is not a miracle but an impossibility.

scrupulous recording, classification and correlation of empirical data, but in endeavouring to give those data a scientific meaning; that is to say, *in proposing causal explanations of them as aspects of material processes and in devising means to test those explanations so that they may become scientific knowledge.*

Firstly, therefore, parapsychology must stop imagining that it is exploring an unknown immaterial world and realize that it is exploring an unknown aspect of the material world; and secondly, it must throw precognition overboard as a nonsensical impossibility not only for scientific materialism but for *any* rational philosophy. It will then at last be on a sound basis to become a science. Although the origins of parapsychology in the West make it unlikely that this will be done, in the USSR the first step, at least, has been taken.

Precognition left behind, no genuine psi-phenomenon is intrinsically incompatible with materialism. Certainly, if science were a closed system, if matter were nothing more than it is known to be, if materialism were the vulgar doctrine that its philosophical opponents set up as a thing of straw for their spiritual shafts, this would be untenable, and both materialism and science would long ago have expired. The contrary being the case, the investigation of psi-phenomena on a scientific basis can only expand scientific knowledge of the world of matter and enrich our materialism.

It would, of course, be foolish to pretend that this is easy, when it is precisely the anomalous nature of psi-phenomena that enables parapsychology to vaunt itself as a spiritual stronghold. The inadequacy of the materialistic hypotheses so far proposed, such as Dr Hans Berger's theory of brain-produced psychical energy,[149] and the development of his resonance concept by Ninian Marshall,[150] not to speak of Dr H. G. Heine's chemical communication system,[151] is clear enough. Nevertheless, the case against the physical nature of psi-energy has been heavily overstated. The repeatedly alleged inapplicability of the inverse square law, for example, so far from established, has actually been contradicted by the outcome of a comprehensive survey of ESP distance experiments: the over-all scoring rates *do* decline with increased distance in approximate accordance with the inverse square law.[152] And even if that law does *not* uniformly apply to psi-transmission, this would still be nothing like a sound argument

149 *The Journal of Parapsychology*, June 1960, pp. 142-148.

150 *The British Journal for the Philosophy of Science*, February 1960.

151 *The Vital Sense*.

152 *The Journal of Parapsychology*, December 1959, Summary of Dr Karlis Osis' paper 'The Distance Problem'.

against its physical nature.[153] The reason why psi is scientifically incongruous is not because it is non-physical, but because its physical nature is not yet known.

Psychical research in the USSR, where scientists prefer to call it 'physiological cybernetics', is at present concentrated on telepathy experiments with hypnotized subjects, the primary aim being to gain control over telepathy as a basis for the investigation of other phenomena, although attention is also being given to such a complicated psi-form as yoga. The working hypothesis is the existence of 'psi-fields' as electromagnetic fields with an additional unknown component, so that theoretically it should be possible to detect psi-processes by their associated electromagnetic effects.[154] Needless to say, these hypothetical 'psi-fields' are in a different category from those of Dr Wassermann's 'field theory of organismic form': they are not positivistic inventions to fill a gap in scientific language, but are conceived as materially existing in the same way as electromagnetic or gravitational fields materially exist, and their supposed existence will be verified or rejected by practical scientific methods.

This is for the future. In the meantime, it is clear that although psi-information may cause sensory hallucinations, it is not received through the senses. This means that the old principle that all information is obtained through the senses must be abandoned. The principle is not an essentially materialist one. It is simply Aristotelian common sense, quite reasonable in the lack of evidence to the contrary, manifestly unreasonable as soon as there is such evidence.

'Extra-sensory perception,' however, is a misnomer; we should be much nearer the mark if we called it 'pre-sensory intuition'. Abundant evidence indicates its extremely primitive nature as the amorphous precursor of sense-consciousness. Its mysterious apprehensions may be sharpened, though seldom into accuracy, by conversion into picture-thinking or even hallucinations, or below the sensory threshold they may arouse the 'feeling of presence', but fundamentally they are totally unconscious.[155]

Their dark vagueness and its easy moulding by suggestion into false certainties make psi-intuitions inevitably unreliable, so that it is distressing to find Rhine proposing ESP as a possible means of long-

[153] Soal & Bateman, op. cit., p. 304; also *Ciba Foundation Symposium on Extra-sensory Perception*, pp. 66, 94.

[154] *Parapsychology Bulletin*, May 1961, November 1961; *The Journal of Parapsychology*, June 1961 (Dr Milan Rýzl, 'Research on Telepathy in Soviet Russia').

[155] Cf. Rhine, *New World of the Mind*, pp. 100-102.

range military intelligence to forestall nuclear attack[156] The grim vision of Pentagon generals in conclave around a professional schizophrenic whose desperate message would bring their heavy fingers hard down upon the push-buttons of apocalypse is not yet actual, but we can never be sure. Who indeed leads us, and whither are we led? . . . 'And the fierce flames burnt round the heavens and round the abodes of men.'[157] But against those who would rather there were total death than the downfall of their 'spiritual values', stands the materialistic affirmation that history has no final synthesis, and that earthborn human life can create higher values than they have ever known.

The world of spirit is a world of shadows. Its phenomena are the illusory projections of an unseen comedy of errors, perhaps not as you like it, but always what you will, whether you believe in measure for measure, all's well that ends well, or rate them much ado about nothing.

That which so far has not been explained is so far inexplicable, and nothing more can now be said of it. *Wovon man nicht sprechen kann, darüber muss man schweigen.*[158] But nothing is finally inexplicable. One by one the obscure strongholds of superstition have crumbled in the advancing light of scientific day; the former universal sway of spiritual causes has been reduced to its last caverns. Unless the primeval night from which humanity is still painfully emerging returns through the threatened self-destruction of our civilization, there need be no doubt that also those last recesses will be illuminated.

And so, to end this our love's labour's lost: '*Satis quod sufficit*. . . . You, that way; we, this way.'[159]

Rome, 1957-1962.

[156] 'Why National Defense Overlooks Parapsychology' (sub-heading 'Concerning Defense Against Physical Attack'), *The Journal of Parapsychology*, December 1957.

[157] Last line of Blake's *America*.

[158] Last proposition of Wittgenstein's *Tractatus*.

[159] *Love's Labour's Lost*, Act 5, first and last lines.

BIBLIOGRAPHY

ALEXANDER, H. G., ed., *The Leibniz-Clarke Correspondence.* Manchester University Press, 1956.

ALEXANDER, S., *Space, Time and Deity.* London, Macmillan, 1920.

AMADOU, R., 'Parapsychology in the Modern Approach to Psychosomatic Man'; v. Wolstenholme & Millar, ed., *The Ciba Foundation Symposium on Extrasensory Perception.*

ANDERSON, M., 'A Precognition Experiment Comparing Time Intervals of a Few Days and One Year', *The Journal of Parapsychology,* June 1959.

and GREGORY, E., 'A Two-Year Program of Tests for Clairvoyance and Precognition with a Class of Public School Pupils'. *The Journal of Parapsychology,* September 1959.

and WHITE, R., 'A Survey of Work on ESP and Teacher-Pupil Attitudes'. *The Journal of Parapsychology,* December 1958.

ARISTOTLE, *Physics.* Oxford, Clarendon Press, 1930.

Poetics. Oxford, Clarendon Press, 1930.

AUGUSTINE, *The Confessions,* tr. F. J. Sheed. London, Sheed & Ward, 1944.

AYER, A. J., *The Problem of Knowledge.* London, Macmillan, 1956.

BERGSON, H., *Matter and Memory,* tr. N. M. Paul and W. S. Palmer, London, Allen & Unwin, 1911.

Time and Free Will, tr. F. L. Pogson. London, Allen & Unwin, 1910.

BERNARDINUS, P., *Institutiones Philosophiae Perennis.* Naples, Cafieri, 1937.

BLAIR, G. W. Scott, *Measurements of Mind and Matter.* London, Dobson, 1950.

BLAKE W., *The Prophetic Writings,* ed. D. J. Sloss and J. P. R. Wallis, Oxford, Clarendon Press, 1926.

BOHM, D., 'A Proposed Explanation of Quantum Theory in Terms of Hidden Variables at a Sub-Quantum Mechanical Level'; v. Körner, ed., *Observation and Interpretation.*

BONDI, H., *Cosmology.* Cambridge University Press, 1952.

BOWDEN, B. V., ed., *Faster than Thought (A Symposium on Digital Computing Machines).* London, Pitman & Sons, 1953.

BRAIN, W. Russell, *Mind, Perception and Science.* Oxford, Blackwell, 1951.

BROAD, C. D. 'Time'. *Encyclopaedia of Religion and Ethics,* Edinburgh, 1921.

The Mind and Its Place in Nature. London, Kegan Paul, Trench, Trubner, 1925.

BROWN, G. Spencer, *Probability and Scientific Inference.* London, Longmans, Green, 1957.

CARINGTON, W., 'Precognition'. *Chambers's Encyclopaedia,* London, 1955.

Chambers's Encyclopaedia. London, George Newnes, 1955.

CHARI, C. T. K., 'Quantum Physics and Parapsychology.' *The Journal of Parapsychology,* September 1956.

CLARKE, W. H. 'A Practical Application of Precognition'. *The Journal of Parapsychology,* December 1957.

CORNFORTH, M., *Science versus Idealism;* two editions. London, Lawrence & Wishart, 1946, 1955.

DESCARTES, R., *Discours de la Méthode*. Paris, Larousse, 1937.
DUNCAN, W. J., *Physical Similarity and Dimensional Analysis*. London, 1953.
DUNNE, G. W., *An Experiment with Time*. London, A. & C. Black, 1927.
The Serial Universe, London, Faber, 1934.
EDDINGTON, A. S., *The Nature of the Physical World*. Cambridge University Press, 1928.
EINSTEIN, A., *The Meaning of Relativity*; 4th edition. London, Methuen, 1950.
and INFELD, L., *The Evolution of Physics*. The Scientific Book Club, London, n.d.
ELIOT, T. S., *Selected Poems*. London, Penguin Books, 1948.
Encyclopaedia Britannica, 14th edition. London, Encyclopaedia Britannica, 1929.
Encyclopaedia of Religion and Ethics, ed. J. Hastings. Edinburgh, T. & T. Clark, 1908-1926.
ENGELS, F., *Anti-Dühring*. Moscow, Foreign Languages Publishing House, 1954.
Dialectics of Nature. Moscow, Foreign Languages Publishing House, 1954.
FEYERABEND, P. K., 'On the Quantum-Theory of Measurement'; v. Körner, ed., *Observation and Interpretation*.
FLEW, A., ed., *Essays in Conceptual Analysis*. London, Macmillan, 1956.
GLASHEEN, A., *A Census of Finnegans Wake*. London, Faber, 1957.
HANSEL, C. E. M., 'A Critical Analysis of the Pearce-Pratt Experiment'. *The Journal of Parapsychology*, June 1961.
HARRIS, Errol E., 'Time and Change'. *Mind*, April 1957.
HEGEL, G. W. F., *The Phenomenology of Mind*, tr. J. B. Baillie, London, Allen & Unwin, 1931.
HEINE, H. G., *The Vital Sense*, London, Cassell, 1960.
HERBST, P., 'The Nature of Facts'; v. Flew, ed., *Essays in Conceptual Analysis*.
HOGBEN, Lancelot, *Mathematics for the Million*, 2nd and 3rd editions. London, Allen & Unwin, 1937, 1951.
HOOKE, S. H., *The Siege Perilous*. London, S. C. M. Press, 1956.
HOYLE, F., *The Black Cloud*. London, Heinemann, 1957.
HUDSON, T. J., *The Law of Psychic Phenomena*. London, G. P. Putnam's Sons, 1893.
HUME, D., *A Treatise of Human Nature*. London, J. M. Dent, 1911.
Enquiry Concerning Human Understanding. Oxford University Press, 1902.
JAMES, W., *The Principles of Psychology*. New York, Dover Publications, 1950.
JEFFREYS, H., *Scientific Inference*. Cambridge University Press, 1957.
JOHNSON, Martin, *Art and Scientific Thought*. London, Faber, 1944.
Science and the Meanings of Truth. London, Faber, 1946.
JOYCE, James, *Finnegans Wake*. London, Faber, 1939.

KENNER, Hugh, *Dublin's Joyce*. London, Chatto & Windus, 1955.
KOOY, J. M. J., 'Space, Time, and Consciousness'. *The Journal of Parapsychology*, December 1957.
KÖRNER, S., 'On Philosophical Arguments in Physics'; v. *Observation and Interpretation*.
ed., *Observation and Interpretation*, Proceedings of the IXth. Symposium of the Colston Research Society. London, Butterworth, 1957.
LEVY, G. R., *The Gate of Horn*. London, Faber, 1948.
LORENZ, K. Z., 'The Role of Gestalt Perception in Animal and Human Behaviour'; v. Whyte, ed., *Aspects of Form*.
MABBOTT, J. D. 'Our Direct Experience of Time'. *Mind*, April, 1951.
'The Specious Present'. *Mind*, July 1955.
MANGAN, G. L., 'An ESP Experiment with Dual-Aspect Targets Involving One Trial a Day'. *The Journal of Parapsychology*, December 1957.
MARSHALL, Ninian, 'ESP and Memory: A Physical Theory'. *The British Journal for the Philosophy of Science*, February 1960.
MASCALL, E. L. *He Who Is*. London, Longmans, Green, 1943.
Christian Theology and Natural Science. London, Longmans, Green, 1956.
MILNE, E. A., *Kinematic Relativity*. Oxford, Clarendon Press, 1948.
MUNDLE, C. W. K. 'How Specious is the Specious Present?' *Mind*, January 1954.
ONIANS, R. B., *The Origins of European Thought*. Cambridge University Press, 1954.
ORWELL, George, *Nineteen Eighty-Four*. London, Secker & Warburg, 1949.
OUSPENSKY, P. D., *In Search of the Miraculous*. New York, Harcourt, Brace, 1949.
Oxford English Dictionary (A New English Dictionary on Historical Principles), ed. James A. H. Murray, Oxford, Clarendon Press, 1888-1933.
POST, H. R., 'Retrospective Miracles or Betting After the Race'. *The British Journal for the Philosophy of Science*, February 1960.
RHINE, J. B., *New World of the Mind*. London, Faber, 1954.
'Why National Defense Overlooks Parapsychology. *The Journal of Parapsychology*, December 1957.
RUSSELL, Bertrand, *Problems of Philosophy*. Oxford University Press, 1912.
'Logical Atomism'; v. *Logic and Knowledge*.
'On Order in Time'; v. *Logic and Knowledge*.
'On the Nature of Acquaintance'; v. *Logic and Knowledge*.
'The Philosophy of Logical Atomism;' v. *Logic and Knowledge*.
Introduction to Mathematical Philosophy. London, Allen & Unwin, 1920.
History of Western Philosophy. London, Allen & Unwin, 1946.
Logic and Knowledge, ed. R. C. Marsh. London, Allen & Unwin, 1956.
RYLE, G., *The Concept of Mind*. London, Hutchinson, 1949.
RYZL, M. 'Research on Telepathy in Soviet Russia'. *The Journal of Parapsychology*, June 1961.

SCHRÖDINGER, E., *Space-Time Structure*. Cambridge University Press, 1950.
SCOTT, C., 'G. Spencer Brown and Probability'. *Journal of the Society for Psychical Research*, June 1958.
SMART, J. J. C. 'The River of Time'; v. Flew, ed., *Essays in Conceptual Analysis*.
SMYTHIES, J. R., *Analysis of Perception*. London, Routledge & Kegan Paul, 1956.
SOAL, S. G. & BATEMAN, F., *Modern Experiments in Telepathy*. London, Faber, 1954.
and BOWDEN, H. T., *The Mind Readers*. London, Faber, 1959.
SÜSSMANN, G., 'An Analysis of Measurement'; v. Körner, ed., *Observation and Interpretation*.
TRAHERNE, T., *Centuries, Poems, and Thanksgivings*, ed. H. M. Margoliouth. Oxford, Clarendon Press, 1958.
UBBELOHDE, A. R., *Time and Thermodynamics*. Oxford University Press, 1947.
VIGIER, J. P., 'The Concept of Probability in the Frame of the Probabilistic and the Causal Interpretation of Quantum Mechanics'; v. Körner, ed., *Observation and Interpretation*.
WALTER, W. Grey, *The Living Brain*. London, Duckworth, 1953.
WARNOCK, G. J., *English Philosophy since 1900*. Oxford University Press, 1958.
WASSERMANN, G. D., 'An Outline of a Field Theory of Organismic Form and Behaviour'; v. Wolstenholme and Millar, ed., *The Ciba Foundation Symposium on Extrasensory Perception*.
WATTS, A. W., *Myth and Ritual in Christianity*. London, Thames & Hudson, 1954.
Webster's New International Dictionary. London, Bell, 1926.
WEST, D. J., *Eleven Lourdes Miracles*. London, Duckworth, 1957.
Psychical Research Today, London, Duckworth, 1954.
WHITROW, G. J., *The Structure of the Universe*. London, Hutchinson, 1949.
The Structure and Evolution of the Universe. London, Hutchinson, 1959.
WHITTAKER, E., *From Euclid to Eddington*. Cambridge University Press, 1949.
WHYTE, L. L., ed., *Aspects of Form*. London, Lund Humphries, 1951.
WITTGENSTEIN, L., *Tractatus Logico-Philosophicus*, 7th impression. London, Routledge & Kegan Paul, 1958.
Philosophical Investigations. Oxford, Blackwell, 1953.
The Blue and Brown Books. Oxford, Blackwell, 1958.
WOLSTENHOLME, G. E. W. & MILLAR, E. C. P., ed., *Ciba Foundation Symposium on Extrasensory Perception*. London, J. & A. Churchill, 1956.

INDEX

For Product Safety Concerns and Information please contact our EU representative GPSR@taylorandfrancis.com
Taylor & Francis Verlag GmbH, Kaufingerstraße 24, 80331 München, Germany

www.ingramcontent.com/pod-product-compliance
Ingram Content Group UK Ltd.
Pitfield, Milton Keynes, MK11 3LW, UK
UKHW020130200926
475702UK00005B/676

* 9 7 8 1 1 3 8 3 9 4 0 2 5 *